GOVERNANCE FOR SUSTAINABLE DEVELOPMENT

Sustainable development stirs up debate about the capacities of political steering and governance. The complexity of the task expounds limits of steering in three dimensions: goals, knowledge, and power. Sustainability goals are subject to changing and controversial risk perceptions, values and interests. Moreover, knowledge of the coupled dynamics of society, technology and nature is limited. Finally, the power to shape structural change in society and technology is distributed across a multitude of actors and societal subsystems. Steering attempts, therefore, have to cope with conflict and ambivalence, with uncertainty, and with a lack of central control; and they have to face the necessity of coordinating different actor groups and social networks.

This volume explores steering strategies and governance arrangements for sustainable development with a view to these problem dimensions. The contributions by authors from various disciplines approach these challenges from different conceptual angles, ranging from positivist, managerial up to post-modern, constructivist perspectives. By combining theoretical reflections with insights from empirical research in European and American contexts, the volume maps out conditions and identifies approaches which both reflect the limits of steering and reveal options for constructively taking up the task of sustainable development in science and practice.

This book was previously published as a special issue of the *Journal of Environmental Policy & Planning*.

Jens Newig is assistant professor at the Institute of Environmental Systems Research, University of Osnabrück, Germany. His main research interests focus on participatory governance, European environmental policy, agenda setting and water policy.

Jan-Peter Voß has been conducting and leading research projects at the Öko-Institut since 2000 and, since 2006, has also become associated research fellow with the Institute for Governance Studies, University of Twente (NL). Prior research centres on governance for sustainable development.

Jochen Monstadt works as a visiting research scholar at the City Institute at York University, Toronto and at the Keston Institute for Infrastructure at University of Southern California, Los Angeles investigating the transition of urban infrastructures in Berlin, Toronto and LA and its impact on urban environmental governance. His research interests include socio-ecological aspects of energy and water systems, infrastructure planning, regional/urban governance and flood risk management.

T0200587

Governance for Sustainable Development

Coping with ambivalence, uncertainty and distributed power

Edited by Jens Newig, Jan-Peter Voß and Jochen Monstadt

Routledge
Taylor & Francis Group

LONDON AND NEW YORK

First published 2008 by Routledge
2 Park Square, Milton Park, Abingdon, Oxon, OX14 4RN

Simultaneously published in the USA and Canada
by Routledge
270 Madison Avenue, New York, NY 10016

Routledge is an imprint of the Taylor & Francis Group, an informa business

Transferred to Digital Printing 2009

© 2008 Edited by Rajah Rasiah, Yuri Sadoi and Rogier Busser

Typeset in Palatino by Techset, Salisbury, UK

British Library Cataloguing in Publication Data
A catalogue record for this book is available from the British Library

ISBN 10: 0-415-45192-2 (hbk)
ISBN 10: 0-415-57220-7 (pbk)

ISBN 13: 978-0-415-45192-5 (hbk)
ISBN 13: 978-0-415-57220-0 (pbk)

CONTENTS

Editorial
Governance for Sustainable Development in the Face of Ambivalence, Uncertainty and Distributed Power: an Introduction

JENS NEWIG, JAN-PETER VOß & JOCHEN MONSTADT

Introduction

Sustainable development has made its way from declaration to practice. New problems have arisen that relate to the question of how to implement sustainable development and through which forms of governance it can be achieved. This is proving to be a challenge. Highlighting the inadequacy of standard approaches of rational problem-solving and steering in the face of ambivalent sustainability goals, uncertain knowledge and distributed steering power, the objective of this issue is to move to the core of these challenges.

'Governance for Sustainable Development' appears to have become a brand name of its own, bundling the research attempts of an abundance of contemporary scholars (see, for a start, the recent collections of papers edited by Brand (2002), Lafferty (2004b), OECD (2002), Petschow *et al.* (2005), Spangenberg & Giljum (2005) or Voß *et al.* (2006)). Perhaps it is no coincidence that broad and encompassing concepts like 'governance' and 'sustainable development' are joined together in this phrase, which has itself been the subject of linguistic enquiry (Farrell *et al.*, 2005). Sustainable development as a highly normative, yet extremely vague concept inescapably calls for a debate on how and with whom, it can be achieved. It raises issues of governance and political steering. Since the Brundtland report and the Rio Conference at the end of the 1980s, much has been achieved in terms of practical approaches and research. Some approaches focus on spatial scales of governance for sustainable development (Moss, 2003; Patterson & Theobald, 1999), such as the European Union (Spangenberg & Giljum, 2005), nation states (Jörgens 2004; Lafferty 2004a; OECD, 2002), cities (Bulkeley & Betsill, 2003) or regional governance (Ostrom *et al.*, 1999). Others highlight particular aspects of governance such as implementation challenges (Bressers, 2004; Lafferty, 2004b; O'Toole Jr., 2004) or participation (Bäckstrand, 2003; 2006; Meadowcroft, 2004; Oels, 2003; Schmitter, 2002; Tippett *et al.*, 2007). All, however, assume to a certain extent that sustainability goals can be assumed as given, focusing on how to put 'it' in place. Often sustainability is equated with environmental protection assuming a basic and/or priority role throughout different policy fields. Sometimes it is equated with 'long term'. In other cases it entails policy integration in the sense that different policies in specific problem orientations are linked, trade-offs become realized and receive explicit political treatment (Scrase & Sheate, 2002).

Research and practical projects started on the basis of well-known approaches to governance. It was assumed that sustainability goals can be defined and operationalized and that one could then evaluate options and find the best way in which to put 'it' in place. Yet, it had increasingly become apparent that sustainable development poses a challenge that is different from other fields of policy. Special characteristics come to light in attempts to conceptualize sustainable development and make it operational for policy (see Lafferty (2004c) on the particular 'differentness' of sustainable development). In this context the uncertainty (see, for example, O'Toole Jr. (2004) and Bressers (2004)) or ambivalence (Mitcham, 1995) becomes highlighted, albeit in a rather cursory manner. Seeking to cope with the complexity and indeterminacy of sustainable development, a recent focus has been placed on socio-technical transitions and system innovations supported by procedural policies to organize societal search process and learning about developmental options and their sustainability (Elzen *et al.*, 2004; Meadowcroft, 2005; Rotmans *et al.*, 2001). In an attempt to capture the conflation of various fundamental issues of modern societal development in debates on sustainable development, Voß & Kemp (2006) have positioned the notion as a 'chiffre', i.e. a cipher by which policy practice reflects and processes basic societal transformations as captured by the concept of reflexive modernization (Beck *et al.*, 1994; 2003).

The conceptual struggle with sustainable development indicates a challenge not only for practical politics, but for the social and political sciences as well. This is illustrated by the emergence of special research programmes (Whitelegg, 2006) and attempts to articulate basic tenets of a new 'sustainability science' (Funtowicz *et al.*, 1998; Kates *et al.*, 2000). An example of new types of research practices induced by the puzzle of sustainable development is a funding programme for social-ecological research by the German Federal Ministry of Education and Research. This programme also made this edition possible by supporting a working group on 'governance and social-ecological transformation' that organized an international workshop from which the papers in this collection are drawn.[1]

The background to this special issue is a specific approach to exploring specificities of governance for sustainable development as complexities that arise from limits to rational steering in three dimensions: goals, knowledge and power (see Voß *et al.*, 2007). Sustainability *goals* are far from being unequivocally defined; rather, the single elements or 'pillars' of sustainability are subject to ongoing controversies based on heterogeneous perceptions, values and interests of individuals and societal groups. Moreover, the *knowledge* of the complex coupled dynamics of society, technology and nature is not only limited, but also regularly contested and highly uncertain. Finally, the *power* to shape structural change in society and technology is distributed across a multitude of actors and societal subsystems. Attempts to bring about societal transformations in terms of a more sustainable development therefore have to cope with conflict and ambivalence, with uncertain and contested knowledge, and with a lack of central control. They also have to face the necessity of co-ordinating different actor groups and social networks.

The papers in this issue explore steering strategies and governance arrangements for sustainable development, focusing on these problem dimensions. Thus, the aim is not to reinvent the steering wheel for sustainable development. The intention is a structured contribution to the discussion of the specific complexity-related traits of governance for sustainability. Papers are presented in five clusters. First, there is an introductory article, followed by three clusters of papers that focus primarily on ambivalence of goals, uncertainty of knowledge and distribution of power, respectively. A final cluster of papers that discuss strategies to cope with complexity in all three dimensions concludes the issue. The papers combine theoretical reflections with empirical studies in a European and North American policy context. Contributions are written from the perspective of various disciplines and conceptual approaches. They also comprise different epistemological stances, ranging from more positivist, managerial approaches to critical, constructivist approaches. In combination, they provide a comprehensive outlook on the specific conditions of governance for sustainable development. The remainder of this editorial provides a brief summary of each paper.

At the outset, **Jan-Peter Voß** *et al.* elaborate on the three dimensions of steering in an introductory article that provides a framework for discussion in other contributions (I). They start from a conventional rationalist ideal of steering based on the unambiguous determination of goals, availability of knowledge to predict consequences and concentration of power to implement strategies. For each of these dimensions they discuss problems that arise in the context of sustainable development. Specific complexities are linked to a combination of the ambivalence of sustainability as a goal, the uncertainty of knowledge due to complex interactions between society, technology and nature, and distributed power to shape structural change in society. With these problem dimensions they construct a typology of steering situations where problems are pronounced differentially. In a second step they present a review of theoretical concepts of societal steering that groups them according to how theories acknowledge ambivalence, uncertainty and distributed power. Based on a typology of problems and a respective review of theories, Voß *et al.* argue for a differentiated discussion of steering capacities with respect to concrete situations, and present an approach to match strategies with problems.

Based on the introductory article, the second section specifically addresses the issue of ambivalence as a problem of governance for sustainable development (II): **Gordon Walker & Elizabeth Shove** do this with a focus on the governance of socio-technical transitions. Drawing on Zygmunt Bauman's works, the authors first contend that ambivalence is not only virtually ubiquitous, but also that the modernist quest for clear definitions and end-states (the 'war against ambivalence') is "futile and counterproductive", only leading to more ambiguity. Postmodern, more reflexive approaches, instead, acknowledge ambiguity as a social essence, encourage us to live with it and make a virtue of necessity, namely to draw on multiple stakeholder perspectives. Secondly, they observe, again following Baumann, that certain concepts are more prone to ambivalence than others, with sustainability clearly belonging to the first category. Drawing on neo-Marxist political ecology, the authors' third main point is to de-mask the strategic rhetoric and power in struggling for definitions of what is 'sustainable' and that this regularly involves both winners and losers. Further commenting on reflexive approaches and taking the Dutch 'transition management' as an example, Walker & Shove criticize these approaches as tending to obscure power play and inequality issues. Finally, the authors advocate a "more thoroughly positive view of ambivalence as a normal rather than a pathological state", permitting and facilitating "forms of movement, change, flexibility and reinterpretation that transitions to sustainability will undoubtedly require".

Bruce Goldstein focuses on the ambivalence of scientific advice and on the incommensurability and competition between epistemic communities. He analyses whether and to what extent alternative scientific advice providing grounds for unanticipated social and ecological consequences can redirect governmental priorities. He explores this by using a case study on wildfire management in California by domestic natural resources agencies, an arena in which an environmental security discourse is thoroughly institutionalized. Specifically, he scrutinizes the attempts of a community initiative of scientists and activists after a large wildfire to promote ecologically sound land-use practices and fostering societal resilience to natural and anthropogenic risks. However, the scientific advice given by the coalition was not recognized as credible knowledge by the government agencies, and the ways to achieve wildfire prevention remained highly contested. He concludes by suggesting that a simultaneous shift of knowledge practice and discursive framework might be a more successful strategy to 'cash in' on the prevailing security discourse, as the recent activities of the national 'Fire Learning Network' indicate.

The three articles of the third section of this edition deal with uncertainty and limited knowledge (III). **Armin Grunwald** gives a synoptic view on the issue of uncertainty in governance for sustainable development. In contrast to contributions in the second part of this edition, which put a focus on the ambivalence of sustainability as a normative orientation, he focuses on problems in the realm of knowledge production that become virulent, irrespective of further complexities resulting from their linkages with diverging perspectives and values. He distinguishes different forms of knowledge and related types of uncertainty. He also argues that sustainable development poses special challenges in that it concatenates and accumulates different types of uncertainties. He infers that working towards sustainable development requires new approaches beyond planning. For this, the openness of the future, indicated by various uncertainties, should be translated into possibilities for shaping development. New approaches will have to allow for learning over time, for adaptation and 'online' modification of measures due to the monitoring of outcomes of governing. For Grunwald, this is a key element of reflexivity in governance that is essential for dealing with the challenge of governing sustainable development.

Hellmuth Lange & Heiko Garrelts address the question as to how the scientific debate on climate change and new scientific uncertainties is being adopted as a basis for political decision making. They analyse how crucial risk issues 'diffuse' into policy. Referring to the debate on blurring boundaries between science and policy making, they discuss the science–policy interface against the background of two empirical case studies on flood protection. In the first case study on coastal flood protection, the administration in charge tries to transform and to curtail the risk issue and its emphasis on uncertainty in a way that makes it compatible with its own safety discourse, thus generating a scientific–administrative hybrid. In the second case—a newly enacted political strategy on riparian flood prevention—policies explicitly draws on uncertainty and risk, thus transferring and integrating the issue firsthand into the political–administrative system. They explain the different risk cultures, referring to the specific governance contexts with their respective institutional setting, actor constellations and situational factors.

To an even greater extent than Lange & Garrelts, **Ineke Meijer & Marko Hekkert** analyse the 'subjective' perceptions of uncertainty by societal actors. Specifically, they analyse the impact of perceived uncertainties of firms on their innovation decisions in contexts of sustainability transitions. The authors distinguish six different sources of uncertainty that can be perceived by business enterprises, including characteristics of the technology at stake and available resources, as well as uncertainty related to the actions of competitors, suppliers, consumers and governments. In their comparative empirical analysis of micro-combined heat-power production and biofuels in the Netherlands, the authors explore the ways in which actors react quite differently to perceived uncertainties, depending not only on the respective policy domain, but also on the actors' experience as entrepreneurs (the more experienced, the more uncertainty inhibits firms from investing) and the phase of the technological transition trajectory (in early phases, uncertainty plays a less crucial role than in later phases). The authors conclude by emphasizing the role of governments and political uncertainties in shaping innovations for sustainable development, and call for "a portfolio of various steering instruments that can be applied in different situations".

The fourth section addresses the problem dimension of distributed power (IV). In a comprehensive account, **James Meadowcroft** investigates this as a key challenge associated with governance for sustainable development. Implications for managing change in a context in which power is distributed across diverse societal subsystems and among many societal actors are discussed. Based on a theoretical analysis of the idea of 'governance for sustainable development', he reflects on the diffusion of power in modern societies. Contrary to conventional theoretical debates, he regards the phenomenon of distributed power not simply as a restriction of effective governance and steering. Instead, he convincingly reveals the potential of distributed power for socio-technological innovation and for the transition of societies towards sustainability and advances some approaches to governing for sustainable development in a radically 'decentred' societal context.

The fifth section concludes this edition by discussing governance approaches that attempt to accommodate complexity in all three problem dimensions of governance (V). A central reference for the three papers is the approach of 'transition management' as developed in the context of sustainable development policy in the Netherlands.

René Kemp, Jan Rotmans & Derk Loorbach reflect on experiences with transition management in the Netherlands. They provide an overview of the model of governance and the process in which it emerged and came to be implemented by the national government to steer developments in energy, transport and agriculture towards sustainable development. A key part of this contribution is the presentation of a framework for the evaluation of resulting policy practices. This emphasizes overcoming political myopia, determination of short-term steps for long-term change and the danger of new lock-ins, as additional points to accommodate ambivalence, uncertainty and distributed power. Emerging governance arrangements in energy provision are scrutinized with the help of this framework. The authors conclude that transition management is applied in different ways from the original model (e.g. established players rather than newcomers play a central role), but that it is still effective in translating the long-term challenge of sustainable development into concrete short-term steps and thereby contributing to changing policy practices and exploiting business interests for structural changes in the economy.

Carolyn Hendriks & John Grin take issue with a family of approaches for governing sustainable development that is denoted as 'reflexive governance'. Reflexive governance arrangements encourage reconsideration of assumptions, institutional arrangements and established practices of governing. This includes strategies to open up institutionalized problem-solving routines and nurture emerging networks that can bring about system innovations for sustainable development. The key point of this paper is to discuss critically, the working of reflexive governance arrangements in practice, especially with regard to the politics that result from their being embedded within broader discursive and institutional contexts. On this basis an alternative conceptualization is offered that explicitly positions reflexive governance as part of a broader discursive system composed of multiple arenas, actors and forms of political communication. This concept is applied to analyse a case taken from Dutch agricultural policy. This reveals a host of struggles as actors try to reconcile the demands of reflexivity (being open, self-critical and creative) with the demands of their existing political world (closed preferences, agenda-driven, control). The authors conclude by outlining implications for understanding politics and legitimacy of reflexive modes of governing for sustainability.

In the final paper in this collection, **Adrian Smith & Andy Stirling** offer a foundational discussion of governance for socio-technical change by referring to concepts presented under headings such as 'transition management' and 'reflexive governance'. They develop two contrasting, ideal-typical conceptualizations of the relations between governance and socio-technical change: 'governance on the outside' and 'governance on the inside'. These types are derived from a discussion of links between social appraisal and social commitment. The 'outside' conceptualization objectifies the subject matter of governance and is managerial in approach; the 'inside' conceptualization is reflexive about how governance co-constitutes its subject matter and is overtly political. The authors discuss how each conceptualization lends itself to contrasting strategies for dealing with uncertainty, ambiguity and power. And they observe how both exist to varying degrees in specific instances of socio-technical governance. The authors argue that understanding the tensions between imperfect attempts to reconcile contradictions between the two is key to understanding dynamics of governance—especially in the area of governance for sustainable development.

Note

1. The papers were presented at an international workshop 'Governance for Sustainable Development' in Berlin, 5–6 February 2006, organized by Jan-Peter Voß, Jens Newig and Jochen Monstadt on behalf of the working group 'Governance and Social-ecological Transformation' (www.sozial-oekologische-forschung.org/de/664.php). All papers have undergone thorough review, both by guest editors, journal editors and two anonymous referees.

References

Bäckstrand, K. (2003) Civic Science for Sustainability: Reframing the Role of Experts, Policy-Makers and Citizens in Environmental Governance, *Global Environmental Politics*, 3(4), pp. 24–41.

Bäckstrand, K. (2006) Democratizing Global Environmental Governance? Stakeholder Democracy after the World Summit on Sustainable Development, *European Journal of International Relations*, 12(4), pp. 467–498.

Beck, U., Bonß, W. & Lau, C. (2003) The Theory of Reflexive Modernization: Problematic, Hypotheses and Research Programme, *Theory, Culture, Society*, 20, pp. 1–33.

Beck, U., Giddens, A. & Lash, S. (Eds) (1994) *Reflexive Modernization. Politics, Tradition and Aesthetics in the Modern Social Order* (Cambridge: Polity Press).

Bressers, H. T. A. (2004) Implementing sustainable development: how to know what works, where, when and how, in: W. M. Lafferty (Ed.) *Governance for Sustainable Development. The Challenge of Adapting Form to Function*, pp. 284–318 (Cheltenham, Northampton: Edward Elgar).

Brand, K.-W. (Ed.) (2002) *Politik der Nachhaltigkeit. Voraussetzungen, Probleme, Chancen - eine kritische Diskussion* (Berlin: edition sigma).

Bulkeley, H. & Betsill, M. M. (2003) *Cities and Climate Change. Urban Sustainability and Global Environmental Governance* (London/New York: Routledge).

Elzen, B., Geels, F. & Green, K. (2004) *System Innovation and the Transition to Sustainability. Theory, Evidence and Policy* (Cheltenham, Northampton: Edward Elgar).

Farrell, K. N., Kemp, R., Hinterberger, F., Rammel, C. & Ziegler, R. (2005) From *for* to governance for sustainable development in Europe: what is at stake for further research?, *International Journal of Sustainable Development*, 8(1/2), pp. 127–150.

Funtowicz, S., Ravetz, J.R. & O'Connor, M. (1998) Challenges in the use of science for sustainable development, *International Journal of Sustainable Development*, 1(1), pp. 99–107.

Jörgens, H. (2004) Governance by diffusion: implementing global norms through cross-national imitation and learning, in W. M. Lafferty (Ed.) *Governance for Sustainable Development. The Challenge of Adapting Form to Function*, pp. 246–283 (Cheltenham, Northampton: Edward Elgar).

Kates, R. W., Clark, W. C., Corell, R., Hall, J. M., Jaeger, C., Lowe, I., McCarthy, J. J., Schellnhuber, H.-J., Bolin, B., Dickson, N. M., Faucheux, S., Gallopin, G. C., Gruebler, A., Huntley, B., Jäger, J., Jodha, N. S., Kasperson, R. E., Mabogunje, A., Matson, P., Mooney, H., Moore, B., O'Riordan, T. & Svedin, U. (2000) Sustainability Science, *Science*, 292, pp. 641–642.

Lafferty, W. M. (2004a) From environmental protection to sustainable development: the challenge of decoupling through sectoral integration, in: W. M. Lafferty (Ed.) *Governance for Sustainable Development. The Challenge of Adapting Form to Function*, pp. 191–220 (Cheltenham, Northampton: Edward Elgar).

Lafferty, W. M. (Ed.) (2004b) *Governance for Sustainable Development. The Challenge of Adapting Form to Function* (Cheltenham, Northampton: Edward Elgar).

xiv *J. Newig* et al.

Lafferty, W. M. (2004c) Introduction: form and function in governance for sustainable development, in: W. M. Lafferty (Ed.) *Governance for Sustainable Development. The Challenge of Adapting Form to Function*, pp. 1–31 (Cheltenham, Northampton: Edward Elgar).

Meadowcroft, J. (2004) Participation and sustainable development: modes of citizen, community and organisational involvement, in: W. M. Lafferty (Ed.) *Governance for Sustainable Development. The Challenge of Adapting Form to Function*, pp. 162–190 (Cheltenham, Northampton: Edward Elgar).

Meadowcroft, J. (2005) Environmental Political Economy, Technological Transitions and the State, *New Political Economy*, 10(4), pp. 479–498.

Mitcham, C. (1995) The Concept of Sustainable Development: its Origins and Ambivalence, *Technology in Society*, 17(3), pp. 311–326.

Moss, T. (2003) Regional Governance and the EU Water Framework Directive: Institutional Fit, Scale and Interplay, in: W. M. Lafferty & M. Narodoslawsky (Eds) *Regional Sustainable Development in Europe. The Challenge of Multi-Level Co-operative Governance*, pp. 207–230 (Oslo: regionet).

OECD (2002) *Governance for Sustainable Development. Five OECD Case Studies* (Paris: OECD).

Oels, A. (2003) *Evaluating stakeholder participation in the transition to sustainable development. Methodology, case studies, policy implications* (Münster: Studien zur internationalen Umweltpolitik).

Ostrom, E., Burger, J., Field, C. B., Norgaard, R. B. & Policansky, D. (1999) Revisiting the Commons: Local Lessons, Global Challenges, *Science*, 284, pp. 278–282.

O'Toole Jr., L. J. (2004) Implementation theory and the challenge of sustainable development: the transformative role of learning, in: W. M. Lafferty (Ed.) *Governance for Sustainable Development. The Challenge of Adapting Form to Function*, pp. 32–60 (Cheltenham, Northampton: Edward Elgar).

Patterson, A. & Theobald, K. S. (1999) Emerging contradictions: sustainable development and the new local governance, in: S. Buckingham-Hatfield & S. Percy (Eds) *Constructing Local Environmental Agendas. People, places and participation*, pp. 156–171 (London, New York: Routledge).

Petschow, U., Rosenau, J. N. & von Weizsäcker, E. U. (2005) *Governance and Sustainability. New Challenges for States, Companies and Civil Society* (Sheffield: Greenleaf).

Rotmans, J., Kemp, R. & Asselt, M.v. (2001) More evolution than revolution: Transition management in public policy, *Foresight*, 3(1), pp. 15–31.

Schmitter, P. C. (2002) Participation in Governance Arrangements: Is there any reason to expect it will achieve "Sustainable and Innovative Policies in a Multi-Level Context"?, in: J. R. Grote & B. Gbikpi (Eds) *Participatory Governance. Political and Societal Implications*, pp. 51–69 (Opladen: Leske & Budrich).

Scrase, J. I. & Sheate, W. R. (2002) Integration and Integrated Approaches to Assessment: What Do They Mean for the Environment?, *Journal of Environmental Policy & Planning*, 4, pp. 275–294.

Spangenberg, J. H. & Giljum, S. (Eds) (2005) Governance for sustainable development, *International Journal of Sustainable Development*, 8(1/2), special issue.

Tippett, J., Handley, J. F. & Ravetz, J. (2007) Meeting the challenges of sustainable development—A conceptual appraisal of a new methodology for participatory ecological planning, *Progress in Planning*, 67, pp. 9–98.

Voß, J.-P., Bauknecht, D. & Kemp, R. (Eds) (2006) *Reflexive Governance for Sustainable Development* (Cheltenham: Edward Elgar).

Voß, J.-P. & Kemp, R. (2006) Sustainability and reflexive governance: introduction, in: J.-P. Voß, D. Bauknecht & R. Kemp (Eds) *Reflexive Governance for Sustainable Development*, pp. 3–28 (Cheltenham: Edward Elgar).

Voß, J.-P. Newig, J., Kastens, B., Monstadt, J., Nolting, B. (2007) Steering for sustainable Development: a Typology of problems and strategies with respect to Ambivalence, Uncertainty and Distributed power, Environmental policy and planning 9(3/4) pp. 185–192.

Whitelegg, K. (2006) The (re)search for solutions: research programmes for sustainable development, in: J.-P. Voß, D. Bauknecht & R. Kemp (Eds) *Reflexive Governance for Sustainable Development*, pp. 273–293 (Cheltenham: Edward Elgar).

Steering for Sustainable Development: a Typology of Problems and Strategies with respect to Ambivalence, Uncertainty and Distributed Power

JAN-PETER VOß, JENS NEWIG, BRITTA KASTENS,
JOCHEN MONSTADT & BENJAMIN NÖLTING

Introduction

Sustainable development, by embracing coupled dynamics of societal and ecological systems on a global scale and over the long term, is an extremely ambitious concept. It provokes dispute because it calls into question concepts, institutions and everyday practices that are based on faith in progress and articulates a responsibility of society for the outcome of these complex interactions. Sustainable development thus stirs up a new debate about the possibility of steering development (Norgaard, 1994). It does so by challenging common assumptions about the definition of goals, predictive knowledge and centralized powers as preconditions of steering.

- Sustainable development concerns the integration or balancing of potentially conflicting values and related risk perceptions. Cultural diversity and ongoing developments make sustainability goals a subject of controversy and change. Steering, therefore, has to cope with ambivalence and conflict.
- Interactions between society, technology and nature lie beyond the reach of disciplinary scientific knowledge. Cause-and-effect relations are highly complex and often non-linear. The predictability of the effects of human intervention is limited. For this reason, steering has to cope with fundamental uncertainty and the possibility of unintended consequences.
- Structural societal changes result from the interplay of diverse factors (political power, law, science, lifestyles, technology, etc.). These factors are not under the control of any one single actor. Many actors with special interests and resources are involved in shaping transformation. On account of this, steering has to face the necessity of co-ordinating the strategies of different actors.

Yet, apart from undermining the foundations of common concepts of steering, sustainable development implies hitherto unknown aspirations to (re-)shape basic social, technological and ecological trajectories of world development. How can these aspirations be lived up to? This fundamental dilemma contained in the concept of sustainable development increasingly becomes taken up in studies on the implications of sustainable development for our understanding of governance (Voß & Kemp, 2006).[1] Here, it will be taken as the starting point of a review and assessment of the current state of thought about societal steering. Specifically, the questions to be asked are 'How do steering problems appear in view of ambivalent goals, uncertain knowledge and distributed power?' and 'How do existing concepts and strategies of steering relate to this new view on problems?'.

In practice, sustainable development is implemented in very diverse contexts, from product development, local agenda processes, and the regulation of hazardous substances to science policy, the transformation of sectoral production and consumption patterns, and the management of global ecological systems (Petschow *et al.*, 2005). All these practical contexts exemplify challenges of ambivalence, uncertainty and distributed power, but to varying degrees. The multitude of steering contexts thus cannot be lumped into a generic steering problem of sustainable development. It is unlikely that there is a single answer to the question 'is it possible to steer sustainable development and, if so, how?'. It seems more appropriate to ask about the particular steering problems that are encountered within concrete empirical contexts. How can these be understood and dealt with?

A typology of problem settings is offered, based upon a differentiation of the dimensions of goals, knowledge and power. This typology can be utilized for a differentiated view on empirical steering contexts as well as for a review of theoretical concepts of steering with respect to what they can offer in terms of strategy recommendations. In effect, it is hoped that this typology allows for deliberation of the match between the problem and the strategy in any particular context of steering for sustainable development.

First, goals, knowledge and power are elaborated as three basic dimensions of steering for sustainable development and criteria are proposed by which the degree of ambivalence, uncertainty and distribution of power within a given context can be assessed. The next step presents a review of selected steering concepts, grouped into five clusters, which scrutinizes their often implicit assumptions on the availability of reliable goals, certain knowledge and centralized power resources. Both typologies are then combined to match empirical problem-settings with theoretical concepts and their strategy recommendations. The concluding section critically discusses the potential of a differentiated approach to the steering problem of sustainable development, as laid out in the previous sections.

Problem Dimensions of Steering for Sustainable Development

In order to provide a anchor for the discussion, the following working definition of steering is used: 'a purposive attempt to bring a system from one state to another by exerting influence on its dynamics of development'. This definition presupposes a subject and an object, but it does not preclude that the steering actor (subject) itself is part of the system to be steered (object). Moreover, it does not imply that steering is actually successful, but makes the relation between intention and outcome a central concern of steering theory and practice.

Steering activities are located in different realms and at different levels of society and can be more or less institutionalized. In any case, parallel steering actions by other actors and within other social domains may interfere with the intentions of a focal actor.[2] The following discussion places a focus on political steering by public actors as prominent addressees of expectations with respect to implementing sustainable development. However, the aim is to present considerations that apply equally to various other forms of steering, e.g. in the context of companies.

Conventional concepts of steering, rooted in a modernist conception of rationality (Stone, 1988), assume that a number of preconditions can be met.

- The *goal*, i.e. the *direction* of the steering attempt is defined clearly and unambiguously.
- Predictive knowledge on system dynamics can be developed, allowing the *effects of action to be assessed*.
- *The power to control* influential factors for system development is given or can be organized.

With regard to steering for sustainable development, these preconditions are called into question.[3] The definition of goals, the analysis of system dynamics and the organization of collective action can become an insurmountable limitation. This may initially cause discomfort and confusion because a framework that has almost become intuitive seems to be lost. At second glance, however, this offers an opportunity to systematically recapitulate concepts of steering and to explore new concepts to possibly arrive at more adequate strategies. It constitutes a starting point for the closer investigation of goals, knowledge and power as problem dimensions of steering for sustainable development.[4] In the following discussion, criteria are proposed that help to locate concrete steering contexts on a continuum between simple, clear, structured (value '0': defined goals, certain knowledge, concentrated power) and complex, heterogeneous, diffuse (value '1': ambivalent goals, uncertain knowledge, distributed power) contexts for each dimension.

The Dimension of Goals: Ambivalence of Sustainability

When it comes to the question of what it means in practice, the concept of sustainable development is affected by its inherent subjectivity and ambiguity (Kates *et al.*, 2005, p. 12). As a consequence, defining sustainability is a normative and political undertaking that basically involves trading off different social goals against each other—even though there is no general currency through which this could be achieved (cf. Becker, 1997, p. 6). Here lies the fundamental ambivalence of goals, which becomes an issue for steering sustainable development (Bauman, 1991; Beck, 2006; Shove & Walker, 2007).

In order to be more precise about the conditions that make sustainability goals within a given steering context more or less ambivalent, two aspects are distinguished: acute conflicts between contending definitions of sustainability goals and the vagueness of agreed sustainability goals.

Conflict of goals.   Since different societal actors weigh values and frame reality differently, they also apply different criteria in appraising options. This is especially relevant to sustainability assessment, since diverse goals that are generally accepted as legitimate, can be achieved simultaneously and to the same extent only seldom. One example is constituted by alternative visions of sustainable energy provision, one based on nuclear energy, the other on renewables. Evaluations of what is an acceptable risk differ greatly between actors and contexts and cannot be determined scientifically (Stirling, 2006). However, even where basic value structures diverge, particular goals can be consensual. Political debate about the depletion of the stratospheric ozone layer, for example, has quickly led to an overarching consensus to phase out the production and use of chlorofluorocarbons. Depending on the degree of conflict within a given problem context—as indicated by statements of politicians, agencies and interest groups—the ambivalence of goals can vary between low (0) and high (1).

Vagueness of goals.   Even those goals that have been determined in a process of collective decision making can remain vague in terms of focus, quantification and time-scale. In order to arrive at a formal political consensus, agreed goals may remain abstract and conceal underlying, unresolved conflicts. Such vague goals may be difficult or impossible to operationalize in practice (Newig, 2007). Much of EU legislation contains this kind of goals. The Water Framework Directive, for instance, prescribes 'a good status' of all European water bodies by 2015. The lack of a clear definition of the term 'good status' leads to disagreement regarding its implementation (Newig *et al.*, 2005).[5] On the other hand, politically defined sustainability goals can also be expressed clearly and unambiguously, so that ambivalence is close to 0. For instance, in November 1990 the German federal government declared a target of reducing the German CO_2 emissions by 25 per cent between 1990 and 2005. This goal was measurable and it was clear when it must be met (whether this was realistic or how it should be implemented is a different question).

The Dimension of Knowledge: Uncertainty About System Dynamics

Steering for sustainability often suffers from the prevalence of uncertain knowledge (Grunwald, 2007). Since the Club of Rome's "Limits to Growth"

(Meadows *et al.*, 1972), along with the methodological criticism it has experienced, there has been a rising awareness that socio-ecological interactions that are relevant to sustainable development are complex in nature and thus difficult to govern (Funtowicz *et al.*, 1998; Vester, 2003).

Sustainable development transgresses traditional knowledge domains. Since many ecological effects of industrial production or certain lifestyles are not perceivable in daily life, science is needed as an intermediary to detect and monitor environmental processes. However, science also often fails to develop a thorough understanding of underlying mechanisms, since many sustainability problems extend well beyond the scope of disciplinarily defined problems and cognitive models that are used to understand them. The issue of BSE/vCJS ('mad cow disease') is one example to portray different kinds of uncertainty. First, the infection of cattle from feeding on animal meal was unforeseen and had to be reconstructed after the phenomenon had appeared. Secondly, the risk of humans coming down with vCJS is still subject to scientific dispute. Thirdly, there is high uncertainty about the effect of different problem solutions, such as import bans or culling (cf. Dressel, 2002). Nevertheless, one should also recognize that not each and every practical steering situation suffers from a high degree of knowledge uncertainty. Two criteria may help to distinguish between different degrees of uncertainty.

Heterogeneity of interacting factors. A problem is considered complex, the more attributes are required to describe it sufficiently. A special aspect of sustainable development problems is that they often comprise interaction between very different elements from the domains of society, technology and nature (Gallopín *et al.*, 2001). Apart from the very different dynamics by which these elements change, it is necessary to interlink knowledge from different domains and disciplines in order to understand how the different dynamics interact and feed into larger socio-ecological patterns. In some steering contexts for sustainable development, however, problems are of a more limited scope and complexity. The shaping or organizational development within a small, hierarchically structured company, for example, requires being able to understand the interaction of more and less homogeneous factors than shaping structural change in national energy systems. The first case would therefore imply lower uncertainty about the effects of steering actions (close to 0) than the second one (close to 1).

Feedback loops and emergent dynamics. Problems are also characterized by the way in which the system is structured, i.e. how elements are interconnected. This refers to the density of interconnections and mutual influences that create feedback loops. Where overlapping feedback cycles are pivotal to system change, dynamics become non-linear, less predictable and exhibit dynamics of self-organization and emergence. Time lags, many indirect effects, (latent) instabilities of systems behaviour, path dependencies and a high sensitivity to initial conditions make intervention a precarious endeavour (Prigogine, 1980; Urry, 2005). Whereas the heterogeneity of interacting factors can, in principle, be dealt with by assembling information and networking different knowledge sources, emergent dynamics imply irreducible, 'radical' uncertainty (Pellizzoni, 2003). Other processes and system contexts of steering for sustainable development are characterized by less complex dynamics. Here, dependent and independent factors can be separated analytically and the outcome of reducing or increasing the value of causing factors can be predicted with satisfactory certainty.

The Dimension of Power: Distribution of Influence to Shape Development

Irrespective of the ambivalence of goals or uncertainty of knowledge, the question of how to implement steering strategies remains a difficulty in itself. Strategies for sustainable development involve deep and radical changes to existing socio-technical structures. Inducing and shaping such changes requires reconfiguring many different factors and processes. The capacities to enact strategies for struc-tural change lie with a broad range of heterogeneous actors who have different problem perceptions, values, ideas and interests about what to do (Kooiman, 1993a; Pierre, 2000a). Co-ordinating activities across these diverse fields is what would be needed to implement strategies to shape societal development (Elzen *et al.*, 2004). Such a view on the sources of influence on societal development sheds light on the different places at which steering for sustainable development has to be located and co-ordinated (Meadowcroft, 2007).

The distribution of power is of special relevance in the context of steering when structural transformation at the level of whole sectors of society comes into focus. Overarching institutional arrangements and competencies for control-ling the diversity of activities that play a role in influencing development are not established in these contexts. For a simplified analytical classification of steering contexts one can distinguish between the horizontal distribution of power along different realms and functional subsystems and the vertical distribution of power along different levels of government.

Horizontal distribution of power.   One example of a case in which the horizontal distribution of power is high (close to 1) is the development of a long-term energy strategy. Many actors from different realms of society need to work together to have an effect (e.g. several ministries, utilities, technology providers, environmental and consumer associations and research institutes). No actor has a role that is hierarchically superior in the sense that it can determine unilaterally the actions of others. One example of horizontally concentrated power (low distri-bution, i.e. close to 0) would be the realization of an eco-management and audit scheme in a small company. Even if the eco-audit must be co-ordinated between different members of staff, the management may have considerable power resources to co-ordinate the implementation process. Indicators for the degree of horizontal distribution of power are the number of policy fields, differ-ent organizations, or actor groups involved in the implementation of steering strategies in combination with differences in power resources that they can dispose of.

Vertical distribution of power.   Vertical distribution of power refers to interdepen-dencies of steering activities at different levels of governance, such as policy making at the level of the European Union, national governments, regional states and local municipalities—or the co-ordination of business strategies at the level of general industrial associations, sector-specific associations, or at the level of single companies and departments within a company. Steering can take place in contexts for which framework regulations are set by higher levels. Speci-fication and enforcement may take place at lower levels. One example is German policy making on higher education in the context of European standards for mutual recognition of academic degrees, on the one hand, and the competencies of the regional states to set legal frameworks for universities and implementation in the form of curricula by university departments, on the other hand. This case shows a high degree of vertical distribution of power (close to 1). On the other hand, there are contexts in which steering occurs more or less independently of higher or lower levels of governance. Land-use planning on a municipal level, for example, is usually implemented relatively free of interference from other governance levels. These cases may be classified as being low in the vertical distribution of power (close to 0). An indicator for the vertical distribution of power is thus the number of different policy or management levels that are involved in implementation and the relative influence that they have on the outcome of steering.

A Typology of Steering Problems

On the basis of the foregoing discussion one can conceive different ideal settings of steering in which ambivalence of goals, uncertainty of knowledge and distribution of power combine in different ways. Table 1 provides a preliminary outline of a typology of steering problems that can be derived from it.

Existing Concepts of Steering

Turning towards concepts and theories of steering, there are different approaches to understand how steering attempts interact with ongoing developments. Consequently, chances of success are assessed differently and a broad range of recommendations are given to steering actors. We are particularly interested in how problems of ambivalence, uncertainty and distributed control are accounted for. To this end, a selection of much-cited steering concepts are reviewed with respect to their implicit or explicit assumptions about goals, knowledge and power. The concepts in the sample differ remarkably in this respect. They are presented in the form of five clusters with a different profile of assumptions:

- steering through regulatory power: crafting rules;
- steering as a problem of goals: providing vision;
- steering as a knowledge problem: learning approaches;
- steering as a power problem: negotiation in networks;
- modulating co-evolution: reflexive governance.

It should be noted that the characterization of these five clusters represents steering theories in an ideal-typical manner. Actual theoretical writings are often more multifaceted. To an even greater extent the practice of political steering and management combines various approaches.

Steering Through Regulatory Power: Crafting Rules

The first cluster of steering concepts does not expound the problems of the issues of ambivalence, uncertainty and distributed power in any explicit manner. Defining goals, assessing effects and implementing strategies is nominated simply as a requirement that has to be met by skilful managers and policy

Table 1. Typology of steering problems

Type of steering problem	Ambivalence of goals	Uncertainty of knowledge	Distribution of power	Description
Full control	Low (0)	Low (0)	Low (0)	Complexity is low in all steering dimensions. Goals can be defined unequivocally, knowledge allows for safe prediction of action consequences, and the actor disposes over power to control all relevant factors to influence system change (example: management decision to purchase energy-efficient office equipment at company level)
Value conflict	High (1)	Low (0)	Low (0)	Knowledge about developmental dynamics and effects of action is available, power to bring about relevant system changes as well, but goals are highly ambivalent, the general direction that development should take is disputed (example: decommissioning of nuclear power plants)
Blind Goliath	Low (0)	High (1)	Low (0)	Goals are defined clearly, power is concentrated highly, but there is very little knowledge on how the system works and what the effects of alternative action options will be (example: natural park administration manages their territory so as to maintain ecosystem stability)
Collective action problem	Low (0)	Low (0)	High (1)	Goals are clear and undisputed, it is known how things work and what would be the effects of alternative actions, but the power to make a change is distributed across many actors that cannot be co-ordinated easily (example: commuter seeks to avoid getting stuck on congested roads)
Disorientated power	High (1)	High (1)	Low (0)	Power to take the system to any future state is all in one hand, but it is unclear what such a desired state would be, nor how the system works and what different actions there would be as a consequence (example: a moronic dictator issues decrees and spends money arbitrarily)
Utopia	Low (0)	High (1)	High (1)	The desired system state can be defined clearly, but there is no knowledge available that would tell one what to do, nor are power resources at hand to have an actual effect on system development (example: fighting detrimental effects of automobility, such as crash victims, noise, pollution, congestion)

(Table continued)

Table 1. Continued

Type of steering problem	Ambivalence of goals	Uncertainty of knowledge	Distribution of power	Description
Clash of interests	High (1)	Low (0)	High (1)	Knowledge on how the system works and what would happen if certain actions were implemented is available, but different goals for future system development are in conflict and are backed by relevant societal powers (example: extension of public transport)
Awkward drifting	High (1)	High (1)	High (1)	Goals are highly ambivalent, knowledge is uncertain and power to influence is distributed. Yet, it is acknowledged that the current development trajectory is hazardous and has to be changed (example: global policy making on sustainable development)

makers. Two classical examples of this cluster are **command-and-control** regulation and **incentive-based instruments**.

Command-and-control concepts of steering assume a central actor who sets goals, knows how the relevant parts of the system function and disposes of the necessary power to implement strategies that are deemed appropriate for goal achievement. The basic concept of command-and-control in environmental policy is to specify the concrete actions that individual polluters must take to solve a problem (Holzinger *et al.*, 2002, p. 2). The control of system dynamics is to be achieved by specifying the behaviour of its elements, top-down, from a centre of control (Gawel, 1994, p. 9; Holzinger *et al.*, 2002, p. 2).

While command-and-control demands specified behaviour, incentive-based steering concepts employ the decision-making capacities of those whose behaviour is to be modified, relying on the market mechanism for co-ordination (James, 1997, p. 11; Rubik *et al.*, 2000, p. 46). Within the wide variety of instruments (e.g. taxes, quotas, tradable permits), it is assumed that social decision making is a mechanical process that can—at least in its aggregated results—be predicted with the aid of market models. Just as in command-and-control concepts, steering actors are assumed to be powerful enough to influence key parameters, such as institutional rules, according to their goals and strategies. Often, incentive-based steering concepts are nested within command-and-control frameworks (Davies & Mazurek, 1998, cited in Cole & Grossman, 1999, p. 892). However, incentive-based instruments imply a slightly lower predictability on the level of individual behaviour (Jänicke *et al.*, 1999, p. 100). An actor might refuse to comply with an economic instrument and accept material disadvantages if the costs of behavioural change are higher than the financial disadvantages of non-compliance. In this context, economic instruments are more voluntary, they provide possibilities of behavioural choice and can be seen as a system of command, but with less control (Cole & Grossman, 1999, p. 895).

The goals of steering, causes of deviance and effects of measures, as well as the implementation process are assumed to be in the hands of the steering actors in both approaches. The outcomes are assumed to be controllable. Against the background of these concepts, steering deficits are often explained as bungling craft or lacking will, rather than fundamental problems of goal definition, knowledge or power.

Steering as a Problem of Goals: Providing Vision

The second cluster of theories focuses on setting goals for public policy and administration and establishing guiding visions for heterogeneous actors. These approaches argue that this is the best way to overcome difficulties with the implementation of programmes and the co-ordination of heterogeneous actors. Thus, steering actors can use their capacities better and more efficiently by providing clear goals or initiating processes of vision-building rather than prescribing particular ways of behaviour.

Concepts such as **new public management** or **management by objectives** are inspired by business methods transferred to public administration and government (Naschold & Bogumil, 1998). The focus of steering is shifted away from detailed prescriptions of individual behaviour towards output indicators for subunits of a system (profit centres, branches of administration or nation states in the realm of international governance). Goals and performance indicators are expected to provide guidance, allowing the assessment of outputs. Respective subunits are free to choose appropriate measures to reach these goals. For example, in climate policy the definition of quantitative targets is regarded a crucial element to steering, because it facilitates implementation and monitoring. Another example of this approach is provided by national environmental action plans or the 'Sevilla Process' on Integrated Pollution, Prevention and Control in Europe.

Yet, there are approaches that emphasize the importance of goals but that are more sceptical about the effectiveness of authoritatively imposed goals. They recommend participatory planning processes in order to generate **shared visions** among those actors who play a role in shaping system development. Examples include approaches from urban and regional planning where various, partially conflicting demands are issued by interest groups, branches of administration, legal provisions and different policy levels. Against this background, shared or jointly developed visions can provide (i) orientation by outlining the main features and the core of planning for all stakeholders, (ii) co-ordination of heterogeneous actors by bridging contradictions of interests and (iii) motivation for participants and a focus for the planning process (Grin & Grunwald, 2000; Cools *et al.*, 2003; Kuder, 2004). Knowledge on system dynamics is considered sufficient or less relevant for action.

All these steering concepts explicitly address the definition of goals as a key to success. With relation to sustainable development, some authors argue that this is a crucial task because the capacities of steering actors to regulate the behaviour of other agents effectively are severely restricted. Guiding visions are seen as a way to co-ordinate actions and processes in different domains of a differentiated social system. In order to get there, it is necessary to critically assess established visions with respect to sustainability and work towards new ones based on integrated assessments of the long-term consequences of socio-ecological development (Brand & Fürst, 2002; Voß *et al.*, 2006).

Steering as a Knowledge Problem: Learning Approaches

The third cluster of concepts primarily addresses problems of complexity and uncertainty of knowledge. Incrementalist approaches focus on learning via small, unsystematic changes and mutual adaptation of actors. Adaptive management stresses systematic policy experiments to cope with complex and changing environments. Participatory approaches build on incorporating the knowledge of heterogeneous actors and on their potential for self-organized, collective learning.

Disjointed incrementalism and the science of "muddling through" (Lindblom, 1959) explain why rational planning so often fails and propose a concept of how steering can even be effective in the face of fundamental uncertainty. Given conflicting societal goals and values, an incrementalist intervention strategy reduces the stakes in political controversies, allowing us to cope with uncertain causal dynamics by being "intelligently exploratory when linked with sequences of trial and error" (Lindblom, 1979, p. 520). To this end, "feedback from policy trials needs to rapidly reach those with authority to make a change", thus avoiding the "accumulation of unfortunate results" (Lindblom & Woodhouse, 1993, p. 133).

Adaptive management was developed in the context of ecosystem management (Holling, 1978; Walters, 1986) in response to limited understanding of the complex dynamics of these systems (Lee, 1999, p. 3). A key strategy is to implement policies as systematic experiments whose outcomes are monitored and analysed for unanticipated side-effects and fed back into the policy system ("learning while doing"; Lee, 1999). The approach has been applied in several cases world-wide (Gunderson *et al.*, 1995; Holling, 1978; Lee, 1995; Röling & Wagemakers, 1998). The majority of these schemes were implemented and led by a centralized decision-maker, mainly the government (Lee, 1995), hence the distribution of intervention capacities was very low.

In recent times, the **participation** of a broad range of societal actors has been proposed to cope with uncertainty and to enhance the information basis of governance attempts (Heinelt, 2002; Newig *et al.*, 2005; Pellizzoni, 2003). Participatory strategies focus on rational argumentation and open deliberation (Dryzek, 1990; 2000; Habermas, 1984) and are proposed to enable social learning (Cross *et al.*, 2004; Pahl-Wostl & Hare, 2004) as well as to provide interaction spaces for the emergence of innovative solutions (Kappelhoff, 2000). This can be enabled by particular procedural settings such as mediation and dialogue (Yankelovich, 1999).

Steering as a Power Problem: Negotiation in Networks

A fourth cluster of steering concepts focuses on negotiations between actors who hold important power resources to influence system development. Various approaches of **network governance** acknowledge that there may be insurmountable differences in values, world views and interests, but that it is still necessary and possible to achieve agreement for co-ordination and effective collective action. Negotiation generally comprises **arguing** and

bargaining (Saretzki, 1996). Nevertheless, there are concepts which focus more on the adaptation of perceived interests through argumentation and others that put emphasis on the allocation of given costs and benefits through bargaining.

Arguing approaches propose to create networks of actors in order to integrate different problem views and interests. They generally place a focus on understanding and agreement and are thus akin to participatory and deliberative strategies sketched above. Bargaining approaches, on the other hand, propose to enhance exchanges between actors to find compromises. In political and planning studies a shift to co-operative network governance is observed as a result of a diffusion of power from the state to specialized social actors and subsystems (Börzel, 1998; Mayntz, 1996). Co-operative arrangements between public and private actors build on the resources of different actors, exploit societal self-steering potentials and enable social learning (Cross *et al.*, 2004). Bargaining in such diverse network settings involves the reciprocal exchange of commitments on the basis of unquestioned world views and self-interests of the different actors. Bargaining can take on sophisticated forms such as log-rolling (Froman & Cohen, 1970) or issue-linkage (Bennet *et al.*, 1998).

Co-ordination and negotiation-based concepts of steering expound the problems of ambivalent goals and uncertain knowledge as a problem of co-ordination for collective action, but not as substantial ethical or epistemological problems.

Modulating Co-Evolution: Reflexive Governance

A last cluster of steering concepts radically parts with the idea of goal-orientated control and instead focuses on enhancing reflexivity of actors with regard to interdependence and embedding in systemic contexts as well as facilitating mutual adaptation in ongoing processes of societal development. They do not believe in the possibility of resurrectong a strong state nor do they negate that it is possible to take influence, at all. Their orientation is towards the modulation of ongoing processes by making use of their own dynamics that can neither be entirely understood nor controlled (Axelrod & Cohen, 2000; Dunsire, 1993; Rip, 1998).

In this perspective sustainable development can be facilitated by the creation of spaces that allow for the anticipation of co-evolutionary interdependence and enhance the reflexivity of actors with respect to their embedding in broader system contexts and dynamics. Interactions in such spaces that cut across institutionalized competencies and specialized perspectives introduce feedback and allow for the mutual adaptation of strategies in order to avoid detrimental "side-effects" (Voß, 2007; Voß & Kemp, 2006).

A specific concept is "**decentralised context steering**" (*Dezentrale Kontext-steuerung*) (Willke, 1992, pp. 189–192, 335–342; 1999, pp. 232–242; see also Jessop, 1997). It reflects steering problems that occur in the context of functional differentiation of modern societies (Luhmann, 1990; Schimank, 1996): autonomous dynamics and particular functional orientations of social subsystems make them blind to the negative impacts on other subsystems and society as a whole. The concept proposes creating feedback through which subsystems can learn about the externalities they cause. This may stimulate adaptation for securing the system's own functions in the long run.

Another concept is **transition management**, which was developed for steering socio-technological change at the sector level. The concept is based on evolutionary and complexity theories. At its core is the creation of 'transition arenas' as platforms for the development of integrative strategies in interaction of diverse stakeholders, but it also employs elements from other steering orientations. It is presented as

> a deliberate attempt to bring about structural change in a stepwise manner ... utilise existing dynamics and orient these dynamics to transition goals that are chosen by society. The goals and policies to further the goals are ... constantly assessed and periodically adjusted in development rounds (Kemp & Rotmans, 2001).

The concept contains specific methodical provisions to deal with problems of goals, knowledge and power, such as the creation of collective visions, experimentation with portfolios of diverse options and negotiation of strategies among stakeholders (Kemp & Loorbach, 2006).

Even though both approaches refrain from the idea of a central steering actor, they assign a special role to the state (political system) as a moderator and supervisor of this decentralized learning and adaptation.

Comparing Problems and Concepts

At the outset of this paper, it was stated that not only were the theoretical concepts of steering very different but also the practical contexts in which the steering takes place. The diversity of steering concepts is, therefore, not a weakness but an asset for developing society's capacity to shape its own future. The multitude of steering theories contains useful conceptualizations of different facets of societal reality that emerge in different combinations, depending on the practical contexts in which steering for sustainable development takes place.

The typology of steering problems, together with the review of steering concepts, thus allows us to match strategies with contexts. The proposed criteria for assessing the ambivalence of goals, uncertainty of knowledge and distribution of power for a given steering context can be used to identify the type of steering problem that actors face in a particular situation. Due to their assumptions about goals, knowledge and power, some theoretical concepts are more appropriate than others. Table 2 gives an overview of the theoretical concepts that offer particular insights and strategic recommendations for each of the problem dimensions of ambivalence, uncertainty and distributed power. The 'scoring' of each cluster of theoretical concepts in terms of offering insights and strategies that address these dimensions does not necessarily imply that these concepts deliver better strategies and outcomes in practice. Here, we are interested in contributions to meet the fundamental conceptual challenges of steering for sustainable development.

As a last step in the explorative appraisal of existing steering theories, an attempt at stock taking of what is proposed in terms of measures to deal with ambivalence, uncertainty and distributed power in practice will be made. This requires a look across the different theoretical approaches to compile basic measures for coping with the complexities of steering for sustainable development. Some of them appear in several theoretical approaches, others are unique to one specific approach. Table 3 presents those measures. Some measures aim

Table 2. Which kind of steering concept helps for which kind of steering problem?

	Concept									
	Crafting rules		Providing vision		Learning approaches		Negotiation in networks		Reflexive governance	
Problem	Command & control	Economic instruments	Management by objectives	Envisioning	Adaptive management	Disjointed incrementalism	Network governance: Arguing	Network governance: Bargaining	Transition management	Decentralized context steering
Ambivalence of sustainability goals			●	●		○	○	○	●	●
Uncertainty of knowledge about system dynamics		○			●	●	●	○	●	●
Distribution of power to shape development		○				○	●	●	●	●

○ Approach recognizes the problem dimension
● Approach concentrates on problem dimension and proposes strategies

Table 3. Measures to cope with ambivalence, uncertainty and distributed power

Measures to ...	... cope with ambivalence of goals	... cope with uncertainty of knowledge	... cope with distribution of power
... map out complexity	Map diversity of goals; participation of relevant actors	Explorative scenario building	Analyse actor constellation (multi-level, multi-domain)
... harness complexity	Open up political debate for articulation of plural goals	Robust strategies for plurality of possible realities	Calibrate, modulate ongoing power struggle
... accommodate complexity	Follow diversity of solutions; revise assessments regularly	Adaptive strategies, learning by doing, experimentation	Create platform/arena for reflexive alignment of actor strategies
... contain complexity	Political decision (e.g. majority vote)	Make existing (local) knowledge available (involvement of experts and stakeholders); confinement to incremental changes, precaution	Influence institutional framework (incentives)
... overcome complexity	Mediate consensus/ development of shared goals	Integrated knowledge production (transdisciplinarity)	Achieve co-ordination through bargaining and arguing

at mapping out ambivalence, uncertainty or distribution of power; others aim at harnessing, accommodating, pragmatically containing or overcoming complexities of steering. Take the example of ambivalence of goals (e.g. in assessing the sustainability of nuclear energy). While mapping diversity of goals does not reduce ambivalence, but helps to understand it, the opening up of public political debate harnesses ambivalences for keeping up an ongoing collective reflection on diverse valuations and trade-offs. Accommodation of ambivalence can be achieved by following various options in parallel that involve differing trade-offs and regularly opening up debate to revise the resulting portfolio. Finally, ambivalence can be contained pragmatically through political decision in legitimate procedures (e.g. majority vote) or overcome in mediation and deliberation that bring about consensus on goals.

Conclusions

Concluding this paper, this section considers what the discussion has yielded in terms of insights into steering for sustainable development. The conceptualization of goals, knowledge and power as three dimensions of steering has allowed us to sketch a typology of steering problems and relate them to theoretical concepts and strategies. This offers a differentiated and context-related discussion of possibilities and limits of steering for sustainable development, making use of the complementary assets of existing theoretical approaches.

With respect to the general question of how to deal with the steering dilemma of sustainable development, it must be reiterated that sustainable development comprises a diverse set of interlocking activities such as market regulation, business management, administrative reform, social mobilization, scientific innovation, educational reform and the shaping of social attitudes in private communication. Not all steering activities are always confronted with a high degree of ambivalence, uncertainty and distributed power.[6] Moreover, conflicting or competing goals, different explanations or interpretations of system dynamics and heterogeneous actors can also be an asset to deal with sustainability problems because they can increase the variety of potential solutions.

The review of steering theories shows that at least some conceptual work on how to deal with ambivalence, uncertainty and distributed power is available. Some approaches are more pragmatic and straightforward, whereas others are more in-depth, orientated towards interactive, open-ended learning and self-organization. Generally, however, there are options to engage in practical experiments in all different kinds of contexts. Practically orientated research and development could focus on the combination of strategies from different conceptual backgrounds and on the development of concrete methods for implementation. It is hoped that this article will further constructive learning in this direction.

Throughout the article reference has been made to 'empirical' steering situations and contexts. This necessarily refers to the perception of a particular steering actor. What the 'empirical' problem-setting, situation and context of steering are can, of course, be contested by other actors. Such dispute is likely to involve differing world views and interests (e.g. dominating actors may propagate their view as a general consensus in order to generate legitimacy; see Mayntz, 2004). The characterization of a steering context with respect to ambivalence of goals, uncertainty of knowledge and distribution of power is therefore also a political process.

The presented typology will therefore not be able to resolve social dispute about appropriate steering strategies. It will rather shift the debate from the general level of steering theory to the concrete level of problem definition. Such a shift can be fruitful for increasing societal capacities to cope with sustainable development. With proposing criteria for assessing the degree of complexity given in each dimension of steering, it is hoped to offer a first ingredient that can help make intersubjective debate about steering situation and strategies possible. Here, however, more elaborate approaches could follow.

Notes

1. Several aspects of this dilemma have been articulated in many places in the literature on environmental policy and sustainable development (see, for example, Bolin *et al.* (2001), Brand (2002), Elzen *et al.* (2004), Jänicke (1986), Kenny & Meadowcroft (1999), Lafferty (2004), Meadowcroft (2005) and Weber (2005)).

2. Our concept of steering can be related to a broader concept of governance (Kooiman, 1993b; Lindberg *et al.*, 1991; Pierre, 2000b). We see steering activities to be embedded in and part of broader patterns of social regulation that comprise formal and informal institutions, actor networks, discourses and technology. Within these patterns various actors carry out steering activities that interfere with each other and create and reproduce these patterns (Voß, 2007). Governance as patterned societal self-regulation thus is an emergent outcome of distributed steering activities (across levels and domains). At the same time, existing governance structures provide an enabling and constraining context for steering (Czada & Schimank, 2000). This notion of governance includes various types of patterns, such as hierachical governmental control, competitive

markets, discursive settings or bargaining practices. It differs from a narrow denomination of governance as co-operative rule-making in networks of public and private actors.

3. A rationalist conception of steering is likewise called into question with respect to other problems and in various contexts of public policy, management of organizations and planning (Dobuzinskis, 1992; Forester, 1989; Rip, 2006; Stacey, 1996; Stone, 1988). The integrative and long-term character of sustainable development, the scope of addressed changes and its prominence in political debate make steering problems in this area particularly salient.

4. In choosing these three dimensions we build on an analysis of steering problems by three guiding questions: What ought to be? What is? What can be done? Of course, the isolation of other problem features, such as scale etc., would also be possible. We suppose, however, that those aspects relevant for steering can be meaningfully subsumed under goals, knowledge or power.

5. Vagueness also allows for agreement on general principles as a first step. It can thus also be productive in fostering the development of collective action capacities in a gradual manner.

6. The type of problem-setting of a steering context may, of course, also change over time, due to ongoing changes or as a steering process goes through different phases. Therefore, it is relevant to reassess the steering situation repeatedly over the course of extended steering processes.

References

Axelrod, R. & Cohen, M. D. (2000) Harnessing Complexity. Organizational Implications of a Scientific Frontier (New York: Free Press).

Bauman, Z. (1991) *Modernity and Ambivalence* (Ithaca, NY: Cornell University Press).

Beck, U. (2006) Reflexive governance: politics in the global risk society, in: J.-P. Voß, D. Bauknecht & R. Kemp (Eds) *Reflexive Governance for Sustainable Development*, pp. 31–56 (Cheltenham: Edward Elgar).

Becker, B. (1997) Sustainability Assessment: A Review of Values, Concepts, and Methodological Approaches, *Issues in Agriculture*, 10 (Washington D.C.).

Bennet, L. L., Ragland, S. E. & Yolles, P. (1998) Facilitating international agreements through an interconnected game approach: the case of river basin, in: R. Just & S. Netanyahu (Eds) *Conflict and cooperation in transboundary water resources*, pp. 61–85 (Norwell, MA: Kluwer).

Bolin, B., The Social earning Group (2001) *Learning to Manage Global Environmental Risks: A Functional Analysis of Social Responses to Climate Change* (Cambridge, Mass: MIT Press).

Börzel, Tanja (1998) Organizing Babylon – On the different concepts of policy networks, *Public Administration*, 76, pp. 253–273.

Brand, K.-W. (2002) *Politik der Nachhaltigkeit. Voraussetzungen, Probleme, Chancen—eine kritische Diskussion* (Berlin: edition sigma).

Brand, W. & Fürst, V. (2002) Voraussetzungen einer Politik der Nachhaltigkeit—Eine Exploration des Forschungsfeldes, in: K.-W. Brand (Ed.) *Politik der Nachhaltigkeit: Voraussetzungen, Probleme, Chancen—eine kritische Diskussion*, pp. 15–110 (Berlin: edition sigma).

Cole, Daniel H. & Grossman, Peter Z. (1999) When is Command-and-Control efficient? Institutions, Technology, and the Comparative Efficiency of Alternative Regulatory Regimes for Environmental Protection, *Wisconsin Law Review*, 5, pp. 887–938.

Cools, Marion, Fürst, Dietrich & Gnest, Holger (2003) *Parametrische Steuerung. Operationalisierte Zielvorgaben als neuer Steuerungsmodus in der Raumplanung* (Frankfurt/M. u.a.: Peter Lang Verlag).

Cross, R., Laseter, T., Parker, A. & Velasquez, G. (2004) Assessing and Improving Communities of Practice with Organizational Network Analysis. The Network Roundtable at the University of Virginia, Virginia.

Czada, R. & Schimank, U. (2000) Institutionendynamik und politische Institutionengestaltung: Die zwei Gesichter sozialer Ordnungsbildung, in: R. Werle & U. Werle (Eds) *Gesellschaftliche Komplexität und kollektive Handlungsfähigkeit*, pp. 23–43 (Frankfurt a.M., New York: Campus).

Davies, J. C. & Mazurek, J. (1998) *Pollution Control in the United States: Evaluating the System* (Washington: Resources for the Future).

Dobuzinskis, L. (1992) Modernist and postmodernist metaphors of the policy process: Control and stability vs. chaos and reflexive understanding, *Policy Sciences*, 25, pp. 355–380.

Dressel, K. (2002) *BSE—The new Dimension of Uncertainty. The Cultural Politics of Science and Decision-Making* (Berlin: edition sigma).

Dryzek, J. (1990) *Discursive Democracy: Politics, Policy and Political Science* (Cambridge: Cambridge University Press).

Dryzek, J. (2000) *Deliberative Democracy and Beyond: Liberals, Critics, Contestations* (Oxford: Oxford University Press).

Dunsire, A. (1993) Modes of Governance, in: J. Kooiman (Ed.) *Modern Governance. New Government–Society Interactions*, pp. 21–34 (London: Sage).

Elzen, B., Geels, F. W., Hofman, P. & Green, K. (2004) Socio-technical scenarios as a tool for transition policy: an example from the traffic and transport domain, in: B. Elzen, F. W. Geels & K. Green (Eds). *System Innovation and the Transition to Sustainability. Theory, Evidence and Policy*, pp. 251–281 (Cheltenham: Edward Elgar).

Forester, J. F. (1989) *Planning in the face of power* (Berkeley, CA: University of California Press).

Froman, L. A. & Cohen, M. D. (1970) Compromise and logroll: Comparing the efficiency of two bargaining processes, *Behavioral Science*, 30, pp. 180–183.

Funtowicz, S., Ravetz, J. R. & O'Connor, M. (1998) Challenges in the use of science for sustainable development, *International Journal of Sustainable Development*, 1(1), pp. 99–107.

Gallopín, G. C., Funtowicz, S., O'Connor, M. & Ravetz, J. (2001) Science for the 21st century: From social contract to the scientific core, *International Social Science Journal*, 53, pp. 219–229.

Gawel, E. (1994) Umweltallokation durch Ordnungsrecht. Ein Beitrag zur ökonomischen Theorie regulativer Umweltpolitik, *Schriften zur angewandten Wirtschaftsforschung*, 65 (Tübingen).

Grin, J. & Grunwald, A. (2000) *Vision Assessment: Shaping Technology in 21st Century Society. Towards a Repertoire for Technology Assessment* (Berlin: Springer Verlag).

Grunwald, A. (2007) Working Towards Sustainable Development in Face of Uncertainty and Incomplete Knowledge, *Environmental Policy and Planning*, 9(3/4) pp. 193–212.

Gunderson, L., Holling, C. S. & Light, S. S. (1995) *Barriers and Bridges to Renewal of Eco-systems and Institutions* (New York: Columbia University Press).

Habermas, J. (1984) *The Theory of Communicative Action, Volume One: Reason and the Rationalization of Society* (Boston: Beacon Press).

Heinelt, H. (2002) Achieving Sustainable and Innovative Policies through Participatory Governance in a Multi-level Context: Theoretical Issues, in: H. Heinelt, P. Getimis, G. Kafkalas, R. Smith & E. Swyngedouw (Eds) *Participatory Governance in Multi-level Context. Concepts and Experience*, pp. 17–32 (Opladen: Leske + Budrich).

Holling, C. S. (Ed.) (1978) *Adaptive Environmental Assessment and Management* (Chichester: Wiley?).

Holzinger, K., Knill, C. & Schäfer, A. (2002) European Environmental Governance in Transition?, Gemeinschaftsgüter: Recht, Politik und Ökonomie; Preprints aus der Max-Planck-Projektgruppe Recht der Gemeinschaftsgüter, Bd. 9. Bonn.

James, D. (1997) Environmental Incentives: Australian Experience with Economic Instruments for Environmental Management, Environmental Economics Research Paper, No.5. C. b. E. o. Canberra, Australia.

Jänicke, M. (1986) *Staatsversagen. Die Ohnmacht der Politik in der Industriegesellschaft* (München: Piper).

Jänicke, M., Kunig, P. & Stitzel, M. (1999) *Lern- und Arbeitsbuch Umweltpolitik. Politik, Recht und Management des Umweltschutzes in Staat und Unternehmen* (Bonn: Verlag J.H.W. Dietz Nachf).

Jessop, B. (1997) The governance of complexity and the complexity of governance: preliminary remarks on some problems and limits of economic guidance, in: A. Amin & J. Hausner (Eds) *Beyond market and hierarchy: interactive governance and social complexity*, pp. 95–128 (Cheltenham: Edward Elgar).

Kappelhoff, P. (2000) Der Netzwerkansatz als konzeptueller Rahmen für eine Theorie interorganisationaler Netzwerke, in: Jörg Sydow & Arnold Windeler (Eds) *Steuerung von Netzwerken*, pp. 25–57 (Opladen: Westdeutscher Verlag).

Kates, R. W., Parris, T. M. & Leiserowitz, A. A. (2005) What is Sustainable Development? Goals, Indicators, Values, and Practice, *Science and Policy for Sustainable Development*, 47(3), pp. 8–21.

Kemp, R. & Loorbach, D. (2006) Transition management: a reflexive governance approach, in: J.-P. Voß, D. Bauknecht & R. Kemp (Eds) *Reflexive Governance for Sustainable Development*, pp. 103–130 (Cheltenham: Edward Elgar).

Kemp, R. & Rotmans, J. (2001) The Management of the Co-Evolution of Technical, Environmental and Social Systems. Paper presented at the 'International Conference Towards Environmental Innovation Systems', 27 September, Garmisch-Partenkirchen.

Kenny, M. & Meadowcroft, J. (Eds) (1999) *Planning Sustainability* (London, New York: Routledge).

Kooiman, J. (1993a) *Modern Governance. New Government–Society Interactions* (London: Sage).

Kooiman, J. (1993b) Social-political governance: introduction, in: J. Kooiman (Ed.) *Modern Governance. New Government–Society Interactions*, pp. 1–6 (London: Sage).

Kuder, T. (2004) *Nicht ohne: Leitbilder in Städtebau und Planung. Von der Funktionstrennung zur Nutzungsmischung* (Berlin: Leue Verlag).

Lafferty, W. M. (Ed.) (2004) *Governance for sustainable development: the challenge of adapting form to function* (Cheltenham: Edward Elgar).

Lee, K. N. (1995) Deliberately Seeking Sustainability in the .Columbia River Basin, in: L. Gunderson C.S. Holling, S.S. Light; (Eds) *Barriers and Bridges to Renewal of Ecosystems and Institutions*, pp. 214–238 (New York: Columbia University Press).

Lee, K. N. (1999) Appraising Adaptive Management, in: *Ecology and Society*, 2nd edn, Art. 3. Available at http://www.ecologyandsociety.org/vol3/iss2/art3/ (acessed 12 May 2005).

Lindberg, L. N., Campbell, J. L. & Hollingsworth, J. R. (1991) Economic governance and the analysis of structural change in the American economy, in: L. N. Lindberg, J. L. Campbell & J. R. Hollingsworth (Eds) *Governance of the American economy*, pp. 3–34 (Cambridge, Melbourne: Cambridge University Press).

Lindblom, C. E. (1959) The Science of Muddling through, *Public Administration Review*, 19, pp. 79–88.

Lindblom, C. E. (1979) Still muddling, not yet through, *Public Administration Review*, 39 (November/December), pp. 517–526.

Lindblom, C. E. & Woodhouse, E. J. (1993) *The Policy-Making Process*, 3rd edn (Englewood Cliffs, N.J.: Prentice Hall).

Luhmann, N. (1990) *Ökologische Kommunikation. Kann die moderne Gesellschaft sich auf ökologische Gefährdungen einstellen?* (Opladen: Westdeutscher Verlag).

Mayntz, Renate (1996) Politische Steuerung: Aufstieg, Niedergang und Transformation einer Theorie, in: Klaus von Beyme & Claus Offe (Eds) *Politische Theorien in der Ära der Transformation*, pp. 148–168. PVS- Sonderheft 26 (Opladen: Westdeutscher Verlag).

Mayntz, R. (2004) Governance im modernen Staat, in: A. Benz (Ed.) *Governance- Regieren in komplexen Regelsystemen. Eine Einführung*. [= Governance, 1], pp. 65–76 (Wiesbaden: Vs Verlag).

Meadowcroft, J. (2005) Environmental Political Economy, Technological Transitions and the State, *New Political Economy*, 10(4), pp. 479–498.

Meadowcroft, J. (2007) Who is in charge here? Governance for sustainable development in a complex world, *Environmental Policy and Planning*, 9(3/4), pp. 193–212.

Meadows, D. H., Meadows, D. L., Randers, J. & Behrens, W. H. (1972) *The Limits to Growth* (New York: Universe Books).

Naschold, F. & Bogumil, J. (1998) Modernisierung des Staates. New Public Management und Verwaltungsreform, *Grundwissen Politik*, 22 (Opladen).

Newig, J. (2007) Symbolic Environmental Legislation and Societal Self-Deception, *Environmental Politics*, 16(2), pp. 279–299.

Newig, J., Pahl-Wostl, C. & Sigel, K. (2005) The Role of Public Participation in Managing Uncertainty in the Implementation of the Water Framework Directive, *European Environment*, 15(6), pp. 333–343.

Norgaard, R. B. (1994) *Development Betrayed. The End of Progress and a Coevolutionary Revisioning of the Future* (London: Routledge).

Pahl-Wostl, C. & Hare, M. (2004) Processes of Social Learning in Integrated Resources Management, *Journal of Community & Applied Social Psychology*, 14(3), pp. 193–206.

Pellizzoni, L. (2003) Uncertainty and Participatory Democracy, *Environmental Values*, 12(2), pp. 195–224.

Petschow, U., Rosenau, J. N. & von Weizsäcker, E. U. (Eds) (2005) *Governance and Sustainability. New Challenges for States, Companies and Civil Society* (Sheffield: Greenleaf).

Pierre, J. (Ed.) (2000a) *Debating Governance* (Oxford; New York: Oxford University Press Inc.).

Pierre, J. (2000b) Introduction: Understanding Governance, in: J. Pierre (Ed.) *Debating Governance*, pp. 1–12 (Oxford; New York: Oxford University Press Inc.).

Prigogine, I. (1980) *From Being to Becoming—Time and Complexity in Physical Sciences*, (San Francisco: W. H. Freeman).

Rip, A. (1998) The Dancer and the Dance: Steering in/of science and technology, in: A. Rip (Ed.) *Steering and Effectiveness in a Developing Knowledge Society*, pp. 27–50 (Utrecht: Uitgeverij Lemma BV).

Rip, A. (2006) A Co-Evolutionary Approach to Reflexive Governance—And Its Ironies, in: J.-P. Voß, D. Bauknecht & R. Kemp (Eds) *Reflexive Governance for Sustainable Development*, pp. 82–101 (Cheltenham: Edward Elgar).

Röling, N. G. & Wagemakers, M. A. E. (Eds) (1998) *Facilitating Sustainable Agriculture. Participatory learning and adaptive management in times of environmental uncertainty* (Cambridge: Cambridge University Press).

Rubik, F., Hoffmann, E. & Simshäuser, U. (2000) *Innovationen durch die Umweltpolitik—Integrierte Produktpolitik (IPP) in Deutschland*. Vorschläge des IÖW. Endbericht zum Gutachten im Auftrag des BMU (Heidelberg).

Saretzki, Thomas (1996) Wie unterscheiden sich Verhandeln und Argumentieren? Definitionsprobleme, funktionale Bezüge und strukturelle Differenzen von zwei verschiedenen Kommunikationsmodi, in: Volkert v. Prittwitz (Ed.) *Verhandeln und Argumentieren. Dialog, Interessen und Macht in der Umweltpolitik*, pp. 19–39 (Opladen: Leske + Budrich).

Schimank, U. (1996) *Theorien gesellschaftlicher Differenzierung* (Opladen: Leske + Budrich).

Shove, E., Walker, G. (2007) Ambivalence, Sustainability and the Governance of Socio-technical transitions, *Environmental Policy and Planning*, 9(3/4), pp. 193–212.

Stacey, R. D. (1996) Management and science of complexity, *Research-Technology Management*, 39(3), pp. 8–10.

Stirling, A. (2006) Precaution, foresight and sustainability, in: J.-P. Voß, D. Bauknecht & R. Kemp (Eds) *Reflexive Governance for Sustainable Development*, pp. 225–272 (Cheltenham: Edward Elgar).

Stone, D. A. (1988) *Policy paradox and political reason* (New York: Harper Collins).

Urry, J. (2005) The Complexity Turn, *Theory, Culture & Society*, 22(5), pp. 1–14.

Vester, F. (2003) *Die Kunst vernetzt zu denken. Ideen und Werkzeuge für einen neuen Umgang mit Komplexität. Der neue Bericht an den Club of Rome*, (first published in 2002) (Munich: Deutscher Taschenbuch Verlag).

Voß, J.-P. (2007) Designs on Governance. Development of policy instruments and dynamics in governance. PhD thesis. Faculty Management and Governance, University of Twente, Enschede.

Voß, J.-P. (2007) Nebenwirkungen und Nachhaltigkeit: Reflexive Gestaltungsansätze zum Umgang mit sozial-ökologischen Ko-Evolutionsprozessen, in: H. Lange (Ed.) *Nachhaltigkeit als radikaler Wandel: Die Quadratur des Kreises?*, (Verlag für Sozialwissenschaften), in press.

Voß, J.-P. & Kemp, R. (2006) Sustainability and reflexive governance: introduction, in: J.-P. Voß, D. Bauknecht & R. Kemp (Eds) *Reflexive Governance for Sustainable Development*, pp. 3–28 (Cheltenham: Edward Elgar).

Voß, J.-P., Konrad, K. & Truffer, B. (2006) Sustainability Foresight. Reflexive Governance in the Transformation of Utility Systems, in: J.-P. Voß, D. Bauknecht & R. Kemp (Eds) *Reflexive Governance for Sustainable Development*, pp. 162–188 (Cheltenham: Edward Elgar).

Walters, C. (1986) *Adaptive Management of Renewable Resources. Biological Resource Management* (New York: Macmillan).

Weber, M. (2005) What role for politics in the governance of complex innovation systems? New concepts, requirements and processes of an interactive technology policy for sustainability, in: U. Petschow, J. N. Rosenau & E. U. von Weizsäcker (Eds) *Governance and Sustainability. New Challenges for States, Companies and Civil Society*, pp. 100–118 (Sheffield: Greenleaf).

Willke, H. (1992) *Die Ironie des Staates. Grundlinien einer Staatstheorie polyzentrischer Gesellschaft* (Frankfurt am Main: Suhrkamp).

Willke, H. (1999) *Systemtheorie II: Interventionstheorie* (Stuttgart: UTB/ Lucius & Lucius).

Yankelovich, D. (1999) *The magic of dialogue: transforming conflict into cooperation* (London: Brealey).

Ambivalence, Sustainability and the Governance of Socio-Technical Transitions

GORDON WALKER & ELIZABETH SHOVE

Introduction

In Voß *et al.*'s (2006a) recent analysis of the different contexts for the governance of sustainability and what they refer to as "steering" processes, ambivalence is identified as a defining characteristic and as more or less problematic. Given that they argue that sustainability involves inherent "subjectivity and ambiguity" and "divergent values", the goals of governance and steering processes are rarely simply and unequivocally defined (Voß *et al.*, 2006a, p. 6). These observations are themselves not especially novel. They resonate with long-standing interest in the multiple definitions and descriptions of sustainability (Dobson, 1998; Lele, 1991; McNeill, 2000) and with critiques of the term's ambiguity, fuzziness, divergent meanings and 'misuse' (Holland, 2000; Mitcham, 1995; Porritt,

1993; Redclift, 1996). However, in this paper it is argued that the implications emerging from a more careful and thorough analysis of ambivalence—its foundations, manifestations and, in particular, its politics—do raise interesting and challenging questions for various styles of governance that can be applied to the pursuit of sustainability objectives. Specifically, we are interested in understanding how ambivalence plays out for the more systemic, adaptive and reflexive approaches to governance that have been advocated recently for the steering of sustainable socio-technical change. In what ways is ambivalence better incorporated into these approaches to governance when contrasted with more conventional modernist models? How might it still be unwisely problematized, obscured or underplayed?

The analysis draws on the work of Bauman (1991), who provides a wide-ranging theoretical and applied account of the complex relationship between ambivalence, modernity and postmodernity. Ambivalence, he argued, is a normal condition of language, such that the "particularly bitter and relentless war against ambivalence" characteristic of "modern times" (p. 3) and of state-directed "social engineering" has proved futile and counterproductive. He argued that ambivalence is both produced and sustained through working against it. This has important implications for an understanding of both conventional and 'reflexive' approaches to the governance of sustainability. Modernist, typically 'linear' attempts to engineer social and environmental change have focused on the realization of clearly defined 'end states' or goals. In this context, uncertainty or dispute about what these 'end states' might be represents a problem to be contained and minimized. By contrast, approaches to the governance of socio-technical transitions have been influenced by traditions of system thinking, which emphasize the complex and dynamic co-evolution of the social and the technical. Having framed sustainability problems in these terms, proponents of 'reflexive' transition management recognize the need to continually review 'goals', to include multiple stakeholder perspectives and 'live with' rather than always seek to resolve conflicts and differences (Elzen & Wieczorek, 2005; Kemp & Loorbach, 2006; Kemp *et al.*, 1998; Smith *et al.*, 2005). Literatures explicitly dealing with what are conceived as complex, multi-actor, multi-level processes of 'enlightened modernist' or 'non-modern' steering and governance (Rip, 2006) consequently approach sustainability as a specific form of 'problem- framing' rather than as a blueprint end-state (Voß & Kemp, 2006).

Moves of this kind evidently recognize and reposition the ambivalence of sustainability as a normative objective and seek to attend constructively to its practical implications. However, as shall be argued, they do so whilst, first, downplaying the politics and dynamics of power that are involved and, secondly, disguising the ambivalences inherent in the process of defining a system to be 'managed' and in the very idea that 'transition managers' have the capacity to intervene in dynamic and emergent complex systems. In this way the locations and moments of ambivalence associated with reflexive forms of governance are not the same as those that arise from more conventional approaches.

Before considering approaches to governance in more detail, the article begins by evaluating the relevance of Bauman's analysis of ambivalence for the conceptualization and promotion of sustainability and for the strategic and political boundary work this involves.

Ambivalence and Sustainability

For Bauman (1991), ambivalence is intrinsic to language, to the act of naming, classifying and categorizing. It is a language-specific disorder, "the possibility of assigning an object or an event to more than one category ... a failure of the naming (segregating) function that language is meant to perform" (Bauman, 1991, p. 1). Language segregates and separates, creates classes which order the world, defining opposites and symmetries, negatives and positives, including and excluding. It is also dynamic rather than static, temporarily naturalized rather than universal, and the site of contest and struggle over definition, meaning and assignment. Ambivalence is for these reasons intrinsic to the language of sustainability, and to the act of attempting to order and categorize in these terms.

When the idea and concept of sustainable development first appeared in the 1980s, became more forcefully articulated through the Brundtland Report, and emerged into the common and dominant discourses of a range of social actors, new meanings were assigned and new language created (Macnaghten & Urry, 1998; Reboratti, 1999). To sustain, to be sustainable, had existing and extant meaning, but when applied to the concerns of environment and development a new set of normatively loaded qualities were added. 'To sustain' already had positive associations—to be unsustainable implying a degree of failure or lack—but when given the task of articulating and providing for a better environment, a better world, a better future, the normative functionality and significance of this term increased a hundredfold. A process of categorization then followed, of reordering, redefining the world into new binaries—what was, and by opposition, what was not sustainable. These new labels were swiftly mobilized, applied and attached as this new mode of sorting and ordering spread. Academics, activists, policy makers and corporate actors sought to identify sustainable and unsustainable forms of growth, types of technologies, systems of production, patterns of consumption, forms of building and construction and so on. Each act of naming was an act of reordering into new binaries in which distinctions between the good and the bad were re-evaluated, restructured and reapplied. As Evans (1998) commented, who possibly could argue that sustainability was a bad thing? Increasingly laying claim to sustainability was to lay claim to a package of cultural and political positives, equally, to be unsustainable was to be associated with the negative, the outmoded and the past rather than the future.

The ordering function of language, and the cultural and political need to divide the world into the progressive and enlightened and the backward and undesired, made and makes this process both inevitable and irresistible. It was not possible to talk about sustainability, or to define it, without simultaneously unleashing its practical and political application and attachment—and hence its ambivalence. As Bauman observed, however strenuous the effort to specify and delineate, the 'language-specific disorder' that ambivalence constitutes is ever present, a constant companion to the naming/classifying function of language. The more precisely those classes are defined, the more likely it is that entities will overflow class boundaries, and that they will be "slimy" (Douglas, 1966) slipping and sliding between and across categories and thus appearing ambivalent, problematic, unsettled and disordered.

To find ambivalence is therefore normal, it is to be expected and, under a starkly relativist account, is everywhere, a function of the instability and contingency of all language and knowledge (Derrida, 1976). However, it is evidently more substantial and more significant for some categories than others. Bauman (1991, p. 53), for example, considered at some length how the concept of 'stranger' embodies deep ambivalences (in contrast to the more ordered oppositional categories of 'friend' and 'enemy'). When faced with the need or intent to categorize in terms of sustainable/unsustainable, we are also more rather than less likely to encounter ambivalence for a set of interconnected reasons. Being 'sustainable' is defined by multiple rather than singular criteria; in applying the label it is necessary to define and classify what is sustainable now, in respect of future unknown, uncertain and unknowable conditions; the term refers to a global universality whilst being applied in local, particular contexts; and its use entails both active definition and categorization, implying answers to both "what is sustainable", as well as "does the subject fall in or outside of the category" (Dobson, 1998). This is not aberrant and unexpected, but intentional, particularly in the Brundtland Commission's construction of the term—a formulation bringing together environment and development discourses, integrating the economic, social and environmental and encompassing multiple values, scales and contexts. 'Sustainability' was deliberately and purposefully used to disrupt previously distinct discourses and domains and to create a new contested language in which ambivalence was necessarily rife. Like many other discursive categories that frame and enroll—community, empowerment, social capital, environmental justice to name but a few—sustainability has "functional malleability" (Gledhill, 1994). Its vagueness, ambiguity and ambivalence can be seen as its strength rather than its weakness (Reid, 1987): it is there to be contested and struggled with. As Jacobs (1991, p. 60) observed

> many political objectives are of this kind—liberty, social justice and democracy. These concepts have basic meanings and almost everyone is in favour of them, but deep conflicts remain about how they should be understood and what they imply for policy.

The relative novelty of the language of sustainability, and its role as what Stirling (2006, p. 236) called a "transcendent evaluative framework", also contributes to the depth and sustenance of ambivalence. Over the past 25 years, the process of adding sustainable/unsustainable labels to existing and new entities has cut across previous normative categories of good and bad—sometimes reinforcing them, sometimes undermining them, and often creating multiple and multiply ambivalent forms of categorization. For example, modes of travelling previously seen as desirable and attractive, have acquired new identities as 'unsustainable' and as ethically problematic forms of behaviour. As travellers, we are consequently presented with a more contested and conflicting set of normative classifications of our available choices and practices.

The political role and strategic resource of ambivalence is important here. For those categories included in the vocabularies of government and governance and of politically significant cultural actors and mediators, and for those which have normative and/or epistemic value, drawing the line between in and out, and between good and bad, is a strategically significant and important act. Not surprisingly, what Gieryn (1983) called "boundary work" becomes a focus for contestation and for the strategically and politically loaded negotiation of definition and meaning. In the case of science (a similarly ambiguous and ambivalent category), which is the subject of Gieryn's analysis, what is being fought over is epistemic authority and legitimacy: those who lose boundary disputes, who are demarcated as 'non-scientific' and therefore lacking in the proper expertise and authority, lose influence and access to the political and cultural resources gained by science (Gieryn, 1983, p. 784). In the case of sustainability, boundary work also involves a struggle for normative authority and for the cultural legitimacy that being inside rather than outside of the boundary can bring (which itself is clearly not unconnected to boundary work around science). If an innovation, a technology, a strategy, a policy, a plan, a way of thinking becomes categorically 'sustainable', economic, social and political 'goodies' of various forms then flow. As Swyngedouw & Heynen (2003) argued, despite rhetorics of integration and consensus there will always be both winners and losers in the pursuit of sustainability and, for this reason, contests over boundary designation need to be seen as political acts in which power is being exercised.

By way of illustration, some corporate and public sector actors have sought to locate energy from waste incineration (a technology situated within an encompassing socio-technical infrastructure) within the domain of both 'sustainable waste management' and 'sustainable energy' policy. In situating it as part of a progressive and legitimate strategy of waste and energy management they position it as something that evidently deserves the support of government policy, planning departments and investors. Such boundary work has been contested by others who define incineration as a dirty, polluting, 'monster' technology—a designation that requires and at the same time generates alternative interpretations of sustainability and entirely different policies for sustainable waste management (Greenpeace, 2001).

Ambivalence (the potential to flex and mould boundaries), simultaneously feeds from and fuels boundary work, this being work through which ambivalence is itself reproduced. As Eden *et al.* (2006) stressed, boundary work is not a once-over exercise but contingent and contextual. The form of boundary work 'required' can differ depending upon the opposition; it can be problematic to transfer the type of 'work' done (or authority gained) from one context to another and boundaries can shift and become unstable through continual renegotiation. The boundary is thus "more properly seen as a fuzzy zone of negotiation and rhetoric, a grey area which may, moreover, be very different for different issues" (Eden *et al.*, 2006, p. 1063). This is a grey area in which the politics of sustainability is necessarily played out and which can only ever partially and provisionally be coloured in.

Transitions, Reflexive Governance and Ambivalence

What does this account and analysis of ambivalence, and of the ambivalent nature of sustainability as a category of language and site of political contestation, mean and imply for different approaches to the governance of sustainability? How does ambivalence and the politics it sustains and draws from factor within the *process* of attempting to shape societal change in general and, specifically, with respect to the governance of socio-technical transitions?

Established traditions of modern government, of management, intervention, control and regulation are not at face value and in their rhetoric well suited to persistent ambivalence. Bauman (1991, pp. 7, 8) argued that a defining characteristic of modernity, has been its need to assert order, to be semantically precise, determinate and predictable:

> The typically modern practice, the substance of modern politics, of modern intellect, of modern life, is the effort to exterminate ambivalence: an effort to define precisely—and to suppress or eliminate everything that could not or would not be precisely defined.

Designers, managers and engineers attempt to fragment the world into a plethora of manageable problems that can be understood, defined ordered and contained. This, Bauman contended, has been a counterproductive and futile endeavour, only serving to sustain and produce yet more ambivalence, fragmentation producing overflows across and clashes between categories with the result that: "Opacity emerges at the other end of the struggle for transparency. Confusion is born out of the fight for clarity. Contingency is discovered at the place where many fragmentary works of determination meet, clash and intermingle" (Bauman, 1991, p. 13).

For the visions and schemes of 'social engineering', including the modernist projects and grand narratives of the twentieth century, such counter-productivity, he argued, was fatally undermining. Bauman consequently asked whether social engineering, lying in disgrace, has any future at all. Must we now accept that "remaking society by design only makes it worse than it was" (Bauman, 1991, p. 269)? And, if so, what are we to make of the parallel conclusion that refraining from social engineering is not without its consequences—as Bauman (1991, p. 270) observed, standing back suits those who are satisfied with the existing order and is thus "hardly ever politically and socially neutral".

In some ways, the idea that we can actively seek a better, more sustainable world and actively manage moves towards sustainability, represents a reassertion of the intent and capacity to 'socially engineer'. That said, methods of 'engineering' are no longer, or at least not exclusively, conceived of in modernist terms. Most obviously, the category of sustainability has been deliberately deployed and, in a sense, designed to create a space in which previously separate and fragmentary 'works of determination' or domains of attempted problem solving could be more purposefully and constructively brought together and intertwined across policy fields, scientific disciplines, governmental institutions and the like. As such, Stirling (2006) argued that the concept represents a challenge to hegemonic Enlightenment notions of progress and to established discourses of science and technology. In addition, increasingly popular approaches like those of sustainable transition management are firmly rooted in traditions of system thinking which highlight the co-evolution of the social and the technical and which seek to understand processes of emergence, transformation and decay. Informed by such approaches, debate about the governance of sustainability has, in general and in principle, moved away from classic modernist models of societal steering. Rip (2006) referred to "enlightened modernist" and "non-modern steering" in order to capture this evolution and reformulation. What does this potentially significant move entail for the ways in which the inherent ambivalence of sustainability is located and handled?

In order to address this question more must be said about how the 'systems in transition' literature conceptualizes change, intervention and agency. Much of this work makes use of Rip & Kemp's (1998) 'multi-level' model of innovation, which distinguishes between the macro-level of the socio-technical landscape, the meso-level regime and the micro-level niche. The key idea here is that change takes place through processes of co-evolution and mutual adaptation within and between these layers. The multi-level model can therefore be used to describe how new technologies emerge (within more or less protected niches) and how they become 'working' configurations that shape and re-shape the regimes and landscapes they sustain and that are in turn sustained by them. In terms of transition, the core task is to figure out how currently dominant socio-technical regimes might be dislodged and replaced and how new configurations might become mainstream. Following this line of enquiry, commentators have sought to describe system dynamics, often through retrospective analyses of the rise and fall of selected socio-technical systems and regimes (e.g. from sail to steam ships, from horse to car (Geels, 2002), or from coal to gas (Correlje & Verbong, 2004)).

Historical and analytical studies of *systems in transition* are typically distanced (even voyeuristic): there is no assumption that better understanding will *necessarily* enhance the capacity to manage or that individuals and organizations can, might or should steer trajectories towards predefined, normative goals.

This has not stopped others from appropriating aspects of this framework or from grafting (sometimes alien) concepts of agency and intervention on to it (Elzen & Wieczorek, 2005; Kemp & Loorbach, 2006). Proponents of *sustainable transition management* have differing views about what is involved and about the extent to which specific objectives can be laid down and pursued. Is transition management a matter of picking one trajectory or another (Kemp, for example, refers to routes that diverge in the forest), or is it a question of shaping or making niches and paths? (Berkhout *et al.*, 2004, p. 50). Alternatively, is it more about managing critical processes of selection and variation within a broader dynamic of socio-technical evolution? These differences aside, all are alike in supposing that intervention in pursuit of normative goals, like those of sustainability, is possible and potentially effective.

There are correspondingly varied conceptualizations of what deliberate intervention actually entails, yet its necessary complexity, subtlety and multidimensionality is recognized widely. Recommendations for action typically favour multiple methods and tools of intervention; opening processes of governance (rather than government) to diverse actors and knowledges; and explicitly accepting the uncertainties and limitations of science-based understanding. In each of these respects the contrast with traditional reflexes of modernist policy making are clear. Elzen & Wieczorek (2005, p. 655) saw transitions in socio-technical systems as "extremely complex processes" that are necessarily multi-actor, multi-factor and multi-level. Inducing and sustaining transitions consequently requires the use of different paradigms of governance at different times and in different contexts. To be effective at all, strategies have to be informed by interactive stakeholder learning and be adapted continually and dynamically over time. In addition, advocates of 'reflexive governance' for sustainable development (Voß & Kemp, 2006), take account of the dynamics and diversity of globalizing processes and of the range of actors now involved in the organization of daily life. A system orientation when combined with ideas of reflexive governance consequently implies not one moment of intervention, following which managers stand back and await the desired result, but a constant process in which further adjustments are made as environmental conditions change, these changes being, in part, the outcome of previous interventions. Feedback, monitoring and circuits of action and reaction are integral to this overall scheme.

Such approaches acknowledge the existence of ambivalence, as we have discussed it, within the tasks and challenges at hand. First, and as already noted, participatory processes of various forms are widely advocated in order to accommodate the multiple perspectives, interests and understandings of the many actors involved. Such processes are sometimes geared toward the construction of collective and consensual visions of what a more sustainable socio-technical system might entail (thereby reducing ambivalence). However, and as discussed further below, others are deliberately designed to 'open up' rather than 'close down' differences of perspective in order to work *with* ambivalence more purposefully than before (Stirling, 2006; Voß et al., 2006b). Secondly, the continued monitoring, feedback and adjustment that is such a distinctive feature of reflexive governance, is promoted not only as a means of handling the interdependencies and unpredictabilities of systemic change, but also because interpretations of what is sustainable and definitions of 'the system' itself, are inherently unstable, becoming more or less ambivalent and requiring re-specification as wider processes, understandings, knowledges and values evolve. Reflexivity is in this sense necessary to handle both the emergent qualities and emergent ambivalences of sustainability. As outlined above, this implies creatively using the *language* of sustainability to categorize and order a changing world.

To give some examples, the category of 'sustainable food' is contested actively (Aiking & Boer, 2004) and becoming more so with divergent pathways opening up around competing claims about the status of organic, fair trade and local food production systems (Goodman, 2004; Guthman, 2004). Similarly for biofuels, heralded as a sustainable alternative to conventional transport fuels, intense debates are currently drawing attention to the wider social and environmental consequences of promoting crop-based fuel substitutes. Rapid and complex patterns of response by multiple actors dispersed across a globalized supply and production system, are producing what some take to be damaging and 'unsustainable' outcomes (including rainforest and habitat destruction, mono-cropping and the potential for regional food shortages) that demand re-evaluation and rethinking of adopted policy measures (UN-Energy, 2007). The ease with which biofuels are now slipping in and out of the sustainable 'category' provides further demonstration of the work involved in responding to the continually changing character of the always provisional 'entities' that are the subject of sustainable transition management.

The literature on transition management and reflexive governance is thus far from blind to ambivalence, and indeed actively seeks to respond to its occurrence and variability. However, there remains an insufficient analysis of the politics that ambivalence produces and feeds from and an incomplete analysis of the locations at which it is to be found. It is to these matters that the article now turns.

The Obscure Politics of Transition Management

The discussion of the nature of ambivalence drew attention to the political and strategic boundary work which goes on around significant language and to the powerful normative legitimacy which the use of the 'sustainable' category can bring. However, strategizing, struggle and contestation are rarely evident in writings about sustainable transition management which instead present and presume a world characterized by co-operation, collaboration, consensus building and 'post-political' claims of common interest (Swyngedouw, 2006) (one rare exception being Smith et al., 2005, p. 1503). The very idea of deliberate transition management supposes some kind of orienting vision or goal and there is a tendency to assume that such an image already exists or can be defined and shared by a constituency of institutional actors. But there is, of course, a politics and dynamics of power involved here, perhaps disguised and obscure, but none the less inevitably at work.

Under closer scrutiny, key questions arise. When and how are the 'sustainable' goals of transition management subject to critical scrutiny, and by whom? Techniques like those of multi-stakeholder involvement in foresight exercises, or methods of public participation and deliberation are not neutral, nor are they evacuated of power and strategic behaviour (Bickerstaff & Walker, 2005; Stirling, 2006). The inclusion of multiple stakeholders can be reinterpreted as a process of co-option and neutering of dissent, producing deeply problematic tensions for those taking part. Who wins and who loses out as transitions are steered and managed in one direction but not another? Kemp & Loorbach (2006, p. 15) referred to the need for strategies to promote "transitions towards more environmentally and socially benign societal systems". While we might conceive technically agreed measures of environmental status, through what means might a more socially benign societal system be identified? For whom is the system more benign, by whose measure and across what space and scale? As Harvey (1996, p. 6) pointed out, the question of what is or is not 'benign' raises matters of justice and politics that cannot be avoided and that will be evaluated in different ways at different places and times.

In practice, some forms of 'sustainable' socio-technical arrangement *may* feature in relatively unproblematic, relatively consensual visions of the future but others, such as the 'sustainable' nuclear-based energy infrastructure currently advocated in the UK (Lovelock, 2004; Mitchell & Woodman, 2006), will not. Categorizing nuclear energy as 'sustainable' is not a merely maverick move. Like other such designations, it has been strategically and skilfully constructed to advance the interests of those in favour of big technology responses to problems of climate change and energy security, and keen to sustain elements of the nuclear technology regime (skills, capacities, profits etc.). The political intensity that is *built into* debates over nuclear technologies, and the contested strategic boundary work that is involved (as in the incineration example referred to earlier) is such that this bumpy territory can never be levelled even by skilfully crafted participatory decision-making processes.

Furthermore, whilst the ambivalence inherent in the goals and objectives of sustainability might be recognized in more reflexive approaches to transition management, there are other locations at which ambivalence and the politics associated with this are also of significance. Most fundamentally there is a politics to the very processes of defining something to manage (the 'it', or system) and to the implication that there are managers of the 'it' who sit outside 'its' boundaries and who can apply transition management tools including levers, niche-building machinery, and engineering devices from a privileged, knowledgeable and external position (Smith & Stirling, 2007). Defining is a matter of abstracting the 'it' in question—the policy, the goal, the system in transition—from its historical environment unless it is the 'whole' that is somehow at stake. This abstracting is not just a technical matter of analysis but an ambivalent, political, constructed and potentially contested process of problem definition (e.g. see Geels (2006) on hygienic transitions).

In other words, templates for transition have to be seen as political statements that are temporally provisional (given that material forms and practices evolve over time); partially inclusive (given the profusion of actors on the social stage), and unavoidably contingent (given that contexts and conditions are dynamic). There is a politics to the governance of transitions that works with and contributes both to the ambivalence of sustainability as a discursive category and to the playing out of power in two key arenas, in the definition of the 'system' in question, and in specifying modes and moments of intervention.

Concluding Comments

Bauman's argument that ambivalence is normal, familiar and reproduced through its attempted elimination, is persuasive and insightful. In following this line of reasoning, this article has analysed the ways in which sustainability as an ambivalent category of language has been given new normative and politically powerful meaning. It has been shown how this has disrupted and redefined existing designations of good and bad and created a new arena for strategic and political boundary work. Building on these ideas, it has been observed that unlike conventionally linear approaches to governance, which focus on realizing clearly defined goals and which therefore seek to minimize ambivalence, strategies of sustainable transition management are frequently informed by more reflexive modes of governance and by a willingness to embrace certain forms of ambivalence—particularly in relation to goal definition. It has also been argued that such strategies can obscure the politics and dynamics of power involved and routinely gloss over related, potentially contentious and also inherently ambivalent questions about how systems are specified and managed.

All this suggests, to follow Baumann's line of argument, that both modernist and reflexive forms of governance breed ambivalence. To this is added the further conclusion that both do so in significantly different ways. In requiring that goals be set, the former condenses debate and uncertainty around the question of what these end-states should be. In refusing to define end-states once and for all, the latter relocates, but does not *remove*, the politics of sustainability. Despite working with, rather than denying, ambivalence with respect to goal definition, proponents of reflexively governed transition management have so far failed to extend this approach to equally fundamental matters of system definition and intervention.

One implication is that what Rip (2006) called "non-modern steering" could and perhaps should take its non-modernism more strongly to heart. If, as Bauman (1991, p. 233) argued, the modernist "rational–universal world of order and truth would know of no contingency and no ambivalence", the non-modern needs to find ways of engaging with multiple forms of instability and disorder and of doing so across the full range of system specification, problem definition, agenda setting and practical action. Voß *et al.* (2006c, p. 429) represented this challenge as a metaproblem of governance in which knowing when to nurture ambivalence and when to diminish it is part of an "efficacy paradox of complexity" characterized by the simultaneous requirement to maintain openness, flexibility and adaptability and, at the same time, reduce each of these in order to make decisions and retain the ability to act. Defined like this, governance is a matter of identifying different possible combinations of 'opening up' and 'closing down' and of finding some balance between them. Though useful, this account plays down the important point that forms of opening also represent moments of closure (and vice versa). In addition, it supposes that debates, problems and questions can be 'opened' and closed at will. What is missing is a more distributed, and in a sense a more complete recognition of the contingency and ambivalence of sustainability as defined and reproduced through the actions, inactions and interactions of multiple, variously powerful agents.

This argues for an extended and more thoroughly positive view of ambivalence as a normal rather than a pathological state both for individuals faced with changing and uncertain conditions and for societies working with liberal notions of democracy, debate and deliberation. Cunningham-Burley (2006, p. 204) made a similar point when discussing the risks associated with human genetics, "Ambivalence and scepticism can be harnessed as a powerful resource for change, whether through the mobilization of public knowledges or the development of greater reflexivity within scientific institutions". Rather than generating fatalism and disengagement, ambivalence permits and facilitates forms of movement, change, flexibility and reinterpretation that transitions to sustainability will undoubtedly require. As represented here, ambivalence is, in a sense, intrinsic to reflexivity: it is the very stuff of a dynamic, critical and questioning liberal politics—difficult, but preferable to absolutist dogmatism and unquestioning certainty.

If this is so, the critical political challenge is to design forms of governance that foster and sustain ambivalence at the various locations and moments that it appears and emerges over time. Rip (2006) made a move in this direction by imagining what such a system might be like. As he explained:

> it will be important to have grey zones and interstices within existing orders. And actors who create difficulties for existing orders, because they irritate, contest, or like tricksters are just mischievous. A further component of reflexive governance is then to entertain such contesting, or merely lateral actions, also when they might obstruct a particular reflexive governance approach as originally envisaged by the agent (Rip, 2006, p. 93).

These future-orientated thoughts prompt us to reflect, finally and also speculatively, on the limits and possibilities of transitions not in sustainability but in governance. If the necessary grey zones and interstices referred to above do not yet exist, one niggling and somewhat ironic question remains: can current forms of governance be 're-engineered' such that positively generative structures of ambivalence come into being? And, if so, what are the mechanisms and who are the architects involved? With this question we come full circle and, in doing so, briefly catch sight of another much less ambivalent world of inequality and power.

Acknowledgement

The authors are grateful for the constructive comments of both referees and the editors of the journal special issue and for the suggestions and observations of colleagues on the notion and meaning of ambivalence.

References

Aiking, H. & Boer, J. D. (2004) Food sustainability: diverging interpretations, *British Food Journal*, 106(5), pp. 359–365.
Bauman, Z. (1991) *Modernity and Ambivalence* (Ithaca: Cornell University Press).

Berkhout, F., Smith, A. & Stirling, A. (2004) Socio-technical regimes, path dependency and transition contexts, in: B. Elzen, F. Geels & K. Green (Eds) *System Innovation and the Transition to Sustainability: Theory, Evidence and Policy*, pp. 48–75 (Cheltenham: Edward Elgar).

Bickerstaff, K. & Walker, G. P. (2005) Shared visions, unholy alliances: power, governance and deliberative processes in local transport planning, *Urban Studies*, 42(5), pp. 2123–2144

Correlje, A. & Verbong, G. (2004) The transition from coal to gas: radical change of the Dutch gas system, in: B. Elzen, F. Geels & K. Green (Eds) *System Innovation and the Transition to Sustainability: Theory, Evidence and Policy*, pp. 114–134 (Cheltenham: Edward Elgar).

Cunningham-Burley, S. (2006) Public knowledge and public trust, *Community Genetics*, 9(3), pp. 204–210.

Derrida, J. (1976) *Of Grammatology* (Baltimore: John Hopkins Press).

Dobson, A. (1998) *Justice and the Environment* (Oxford: Oxford University Press).

Douglas, M. (1966) *Purity and Danger* (London: Routledge and Keagan Paul).

Eden, S., Donaldson, A. & Walker, G. (2006) Green groups and grey areas: scientific boundary work, NGOs and environmental knowledge, *Environment and Planning A*, 38, pp. 1061–1076.

Elzen, B. & Wieczorek, A. (2005) Transitions towards sustainability through system innovation, *Technological Forecasting and Social Change*, 72, pp. 651–661.

Evans, B. (1998) The rhetoric of Rio and the problem of local sustainability, in: P. T. Kivell, P. Roberts & G. Walker (Eds) *Environment, Planning and Land Use*, pp. 42–56 (Aldershot, Avebury).

Geels, F. (2002) Technological Transitions as evolutionary reconfiguration processes: a multi-level perspective and case study, *Research Policy*, 31(8–9), pp. 1257–1274.

Geels, F. (2006) The hygienic transition from cesspools to sewer systems (1840–1930): The dynamics of regime transformation, *Research Policy*, 35(7), pp. 1069–1082.

Gieryn, T. (1983) Boundary-work and the demarcation of science from non-science: strains and interests in professional ideologies of scientists, *American Sociological Review*, 48, pp. 781–795.

Gledhill, J. (1994) *Power and its disguises* (London: Pluto Press).

Goodman, M. K. (2004) Reading Fair Trade: political ecological imaginary and the moral economy of fair trade foods, *Political Geography*, 23, pp. 891–915.

Greenpeace UK (2001) *Money to Burn – pollution and health impacts of incinerating resources* (London: Greenpeace UK). Available at: http://www.greenpeace.org.uk/media/reports/money-to-burn (1 June 2007).

Guthman, J. (2004) Back to the land: the paradox of organic food standards, *Environment and Planning A*, 36(3), pp. 511–528.

Harvey, D. (1996) *Justice, Nature and the Geography of Difference* (Oxford: Blackwell).

Holland, A. (2000) Introduction—Sustainable Development: The Contested Vision, in: L. Keekok, A. Holland & D. McNeill (Eds) *Global Sustainable Development in the 21st Century*, pp. 1–9 (Edinburgh: Edinburgh University Press).

Jacobs, M. (1991) *The Green Economy* (London: Pluto Press).

Kemp, R. & Loorbach, D. (2006) Transition Management: a reflexive governance approach, in: J.-P. Voss, D. Bauknecht & R. Kemp (Eds) *Reflexive Governance for Sustainable Development*, pp. 103–130 (Cheltenham: Edward Elgar).

Kemp, R., Schot, J. & Hoogma, R. (1998) Regime shifts to sustainability through processes of niche formation: the approach of strategic niche management, *Technology Analysis and Strategic Management*, 10, pp. 175–196.

Lele, S. (1991) Sustainable Development: a critical review, *World Development*, 19(6), pp. 607–621.

Lovelock, J. (2004) Something Nasty in the Greenhouse. Paper presented to the 'GAIA and Global Change Conference', Dartington, Devon, UK, 4 June. Available at http://www.world-nuclear.org/opinion/lovelock_040604.htm (2 October 2007).

Macnaghten, P. & Urry, J. (1998) *Contested Natures* (London: Sage).

McNeill, D. (2000) The Concept of Sustainable Development, in: L. Keekok, A. Holland & D. McNeill (Eds) *Global Sustainable Development in the 21st Century*, pp. 10–30 (Edinburgh: Edinburgh University Press).

Mitcham, C. (1995) The Concept of Sustainable Development: its Origins and Ambivalence, *Technology in Society*, 17(3), pp. 311–326.

Mitchell, C. & Woodman, B. (2006) *New Nuclear Power: implications for a sustainable energy system* (London: Green Alliance).

Porritt, J. (1993) Sustainable Development: panacea, platitude or downright deception, in: G. Cartledge (Ed.) *Energy and Environment* (Oxford: Oxford University Press).

Reboratti, C. E. (1999) Territory, Scale and Sustainable Development, in: E. Becker & T. Jahn (Eds) *Sustainability and the Social Sciences*, pp. 207–222 (London: Zed Books).

Redclift, M. (1996) *Wasted: Counting the Costs of Global Consumption* (London: Earthscan).

Reid, D. (1987) *Sustainable Development: an introduction* (London: Earthscan).

Rip, A. (2006) A co-evolutionary approach to reflexive governance—and its ironies, in: J.-P. Voss, D. Bauknecht & R. Kemp (Eds) *Reflexive Governance for Sustainable Development*, pp. 82–100 (Cheltenham: Edward Elgar).

Rip, A. & Kemp, R. (1998) Technological change, in: S. Rayner & E. Malone (Eds) *Human Choice and Climate Change*, vol. 2 (Columbus, Ohio: Battelle).

Smith, A. & Stirling, A. (2007) Moving outside or inside? Objectification and reflexivity in the governance of socio-technical systems, *Journal of Environmental Policy and Planning*, 9(3,4), pp. 213–225.

Smith, A., Stirling, A. & Berkhout, F. T. (2005) The governance of sustainable socio-technical transitions, *Research Policy*, 34, pp. 1491–1510.

Stirling, A. (2006) Opening Up or Closing Down? Analysis, participation and power in the social appraisal of technology, in: M. Leach, S. Scoones & B. Wynne (Eds) *Science and Citizens: Globalization and the Challenge of Engagement* (London: Zed).

Swyngedouw, E. (2006) *Impossible/Undesirable Sustainability and the Post-Political Condition*. Paper presented at the Royal Geographical Society with the Institute of British Geographers Conference, London, September.

Swyngedouw, E. & Heynen, N. (2003) Urban political ecology, justice and the politics of scale, *Antipode*, 35(5), pp. 898–918.

UN-Energy (2007) Sustainable Bioenergy: a framework for decision makers, United Nations. Available at http://esa.un.org/un-energy/pdf/susdev.Biofuels.FAO.pdf (1 June 2007).

Voß, J.-P. & Kemp, R. (2006) Sustainability and Reflexive Governance: Introduction, in: J.-P. Voss, D. Bauknecht & R. Kemp (Eds) *Reflexive Governance for Sustainable Development*, pp. 3–28 (Cheltenham: Edward Elgar).

Voß, J.-P., Kemp, R. & Bauchnecht, D. (2006c) Reflexive Governance: a view on an emerging path, in: J.-P. Voss, D. Bauknecht & R. Kemp (Eds) *Reflexive Governance for Sustainable Development*, pp. 419–437 (Cheltenham: Edward Elgar).

Voß, J.-P., Newig, J., Nolting, B., Kastens, B. & Monstadt, J. (2006a) Steering for Sustainable Development – Contexts of Ambivalence, Uncertainty and Distributed Power. Paper prepared for workshop on 'Governance for Sustainable Development: Steering in the Context of Ambivalence, Uncertainty and Distributed control', mcBerlin, 6–7 February.

Voß, J.-P., Truffer, B. & Konrad, K. (2006b) Sustainability foresight: reflexive governance in the transformation of utility systems, in: J-P. Voss, D. Bauknecht & R. Kemp (Eds) *Reflexive Governance for Sustainable Development*, pp. 162–188 (Cheltenham: Edward Elgar).

The Futility of Reason: Incommensurable Differences Between Sustainability Narratives in the Aftermath of the 2003 San Diego Cedar Fire

BRUCE EVAN GOLDSTEIN

Introduction

From its origins in progressive-era forestry in the early twentieth century (O'Riordan, 1988), the idea of sustainability has been adopted across a wide range of planning and policy arenas to identify how humans should organize themselves and relate to their environment. With the diffusion of sustainability, the idea has been both praised and criticized for having many and contradictory meanings (Newton & Freyfogle, 2005; Redclift, 2006; Williams & Millington, 2004). The editors of this special issue (Voß *et al.*, 2007) suggest that this range of meanings reflects a distribution of values and risks across a range of social objectives. Walker & Shove (2007) argue that this diversity is not only inevitable but also inescapable, since efforts to reconcile its multiple meanings fail because the language used to describe sustainability is both unstable and contingent.

This multiplicity of definitions has nurtured more than a decade of intellectual fecundity in which sustainability has served as the "mantra that launched a thousand conferences",[1] including the one that sponsored this collection of papers. Scholarship has thrived in part because a diversity of meanings is so unacceptable to those with a passionate interest in promoting adoption of a particular definition of sustainability. One prominent example of this has played out within the field of international affairs. First, a group of scholars, journalists and activists proposed that sustainable management of the environment and natural resources was a prerequisite for national security. Environmental degradation and scarcity, they argued, were linked directly to the destabilizing flow of environmental refugees and struggles between states to secure natural resources, whether timber, oil or water (Homer-Dixon, 1999; Kaplan, 2000).

While this coupling of sustainability and security succeeded in convincing some Clinton-era policy elites to support what Prins (2004) called the "security bonus" of enhanced governmental commitment and resources for environmental programs, the effort was also criticized for pandering to the interests and perspectives of the powerful (Barnett, 2001; Dalby, 2002). Dalby and his fellow critics suggested that linking environmental programs to national security concerns might yield the honest trifles of state patronage, only to betray efforts to realign human–environmental relationships in deepest consequence by re-inscribing exploitative neocolonial geopolitics, instrumental conceptions of nature and Western consumerism. Their critique focused on theoretical fallacies and factual errors of interpretation, particularly in relation to the idea that environmental deterioration would drive Third-World anarchy across First-World borders. According to Dalby and his allies, concerns about this supposed threat were not supported by the historical record of conflict dynamics or data on each nation's reliance on global resource flows. Drawing support from history, economics, ecology and political science, these critics sought to demonstrate that the only way to achieve security was to displace the referent of sustainability from the Western consumerist state to the encompassing biosphere.

This debate highlighted two approaches to sustainability, one couched in the language and assumptions of the politically powerful, the other drawing on a myriad of intellectual disciplines to challenge those assumptions and frame a transformative alternative. Tactically, the choice between these two approaches to advocating sustainability hinges on the faith that one has in the capacity and willingness of others to consider evidence that calls their existing commitments into question, and then realign with an alternative vision of sustainability. Can factual and theoretical arguments narrow the differences between adherents of different approaches to sustainability, or even help them to appreciate the basis of these differences? Is convincing others to adopt a new definition of sustainability just a matter of overcoming ignorance, entrenched interests and bias, or are there other reasons why reason is unable to overcome the ambivalence of sustainability?

This paper addresses these questions by describing an intense and intimate engagement between proponents of two definitions of sustainability, each quite similar to the alternatives proposed within international affairs. In the aftermath of the 2003 Cedar fire in San Diego, the largest wildfire in California over the past century, a coalition of scientists and activists developed a conception of sustainability that was akin to Dalby's ideas about the need to sustain the biosphere to ensure security. This coalition was opposed to the policies of regional natural resource agencies, who had long garnered the 'security bonus' embedded in a progressive-era[2] narrative of sustainable resource conservation. The scientists and activists attempted to convince the natural resource agencies to adopt policies compatible with the coalition's conception of sustainability by deploying what they regarded as sound and compelling scientific arguments.

The results were not what the coalition had hoped. Far from welcoming their advice, the natural resource agencies rejected the coalition's scientific arguments and acted swiftly to silence the community-based initiative. Coalition members attributed their failure to the agencies' scientific ignorance and refusal to acknowledge truths that threatened the established order. A more symmetrical approach to understanding their frustrated efforts is taken here, by placing the agencies' seemingly intolerant reaction to coalition science in the context of the long-standing reliance of San Diego's government agencies on other forms of knowledge and expertise. Pitted against these long-established ways of knowing, the community coalition's scientific expertise was recognized neither as authoritative nor relevant to achieving agency objectives. Quite the contrary—what was compelling knowledge to the coalition activists was not only inaccurate to the agencies, it also threatened their capacity to pursue their own ideas about sustainability.

In this way, the article draws on ideas about the co-production of knowledge and the social order (Jasanoff, 2004) to suggest that the rejection of coalition claims by the natural resource agencies was not motivated merely by bias and vested interest. For each side in this conflict, the manner in which knowledge was co-produced is traced, along with the other institutional commitments that constituted their respective sustainability discourses. Complementing Walker & Shove's (2007) concern with the irreducible ambiguity of the language of sustainability, it is concluded that conceptions of sustainability may be incommensurable because they are informed and justified by different knowledge practices. If knowledge practices not only underpin associated conceptions of sustainability but are also co-produced with them, knowledge is both a driver and an outcome of the efforts of particular actors to achieve a particular conception of sustainability. This dialectic of knowledge and the social order precludes the possibility of a universally valid science that can adjudicate between contesting sustainabilities.

Incommensurability

Incommensurability was defined by Kuhn (1970) as part of his questioning of the existence of a single scientific community and the continuity of scientific progress. Kuhn began by identifying the basis of epistemological pluralism among the different scientific disciplines. In each discipline, unique research methods, model experiments, and technical languages served to define which questions were significant and prefigure the appropriate answers. Communication between communities, let alone collaboration, was hampered by these methods, experiments and languages because they could not be acquired easily, since they were learned through practice rather than explicit formulation and constituted a kind of 'craft knowledge'. Even specialists within disciplines were unable to agree with their scientific colleagues during times when their discipline was undergoing a 'paradigm shift' in response to new ideas and findings. Kuhn held that the disciplines could not be integrated sensibly because science was not a seamless whole.

This insight applied both between scientific disciplines and through time within scientific disciplines, since introduction of a new paradigm created incommensurability, the impossibility of thinking back into what preceded it.

Kuhn's ideas had profound implications for understanding enduring social differences, since incommensurability suggests that knowledge is not independent of the particularities of how it was produced and that ways of coming to knowledge cannot be collapsed in accordance with a single and universal logic. However, Kuhn did not consider how scientific communities engaged with the broader society and culture or examine the ways that non-scientific knowledge communities functioned (Fuller, 2000). Because of this circumspection, Kuhn's ideas about scientific knowledge do not constitute an overt challenge to the predominant instrumental view of science within planning and policy making, a view that science is useful to political stakeholders only in order to rationalize support policies previously arrived at through political calculation (Flyvbjerg, 1998; Majone, 1989). For example, a powerful political faction could use expertise as a means to close policy debate by turning discussion from desired ends to efficient means (Ellul, 1964), although the flow of events can produce contingencies that allow policy entrepreneurs to identify appropriate problems and apply pre-packaged solutions whose virtues are demonstrated by credible expertise (Kingdon, 1984). Within this tradition, while scientific truths may be politically convenient or inconvenient they are not constitutive of policy and perspectives, nor are these truths shaped by social dynamics outside of science. Social dynamics can only degrade scientific truths by introducing a source of bias.

Anthropological studies of the interrelationships between traditional or place-based knowledge and cultural identity were first to breach this firewall separating knowledge practices from culture and society. This appreciation for heterogeneous knowledge practices was accompanied by analysis of how peripheral communities could be deprived of the benefits of their own expertise as well as their cultural integrity when central governments exercised power in the name of scientific rationality (Scott, 1998; Wynne, 1996). However, these studies that established the relationship between traditional knowledge and maintenance of a traditional social order did not always apply this analytical framework symmetrically to scientific knowledge. For example, Coburn (2003) suggested that science is distinguished from local knowledge precisely by the possession of invariant characteristics, such as a commitment to falsifiability (Popper, 1959). Scott (1998, p. 331) suggested a functional explanation for this categorical distinction between local and scientific knowledge, since "High modernism has needed this 'other', this dark twin, in order to rhetorically present itself as the antidote to backwardness".

In the 1970s and 1980s, science studies researchers questioned this epistemological privilege claimed by science over any other forms of knowledge and began to integrate scientific practice within a broader matrix of social practices, institutional context and cultural norms (Hess, 1997). Inspired by Foucault's (1977) ideas about the inextricable relationship between knowledge and power, analysts began examining all forms of knowledge as "situated" (Haraway, 1996), both individually in relation to a perspective, position and embodiment and collectively in terms of governance and cultural expression. As one seminal co-productionist study concluded, "Solutions to the problem of knowledge are solutions to the problem of social order" (Shapin & Schaffer, 1985, p. 332). Early work concerned with public policy and the environment examined how scientific truths were shaped by controversy and contestation, such as Jasanoff's (1987) work on how regulatory scientists rely heavily on statistical evidence in order to endure legal scrutiny. More recent studies have examined how science is situated in particular settings, from the co-production of regulatory science and policy in different nations (Jasanoff, 2005) to new knowledge practices within the emergent regulatory agencies of the European Union (Waterton & Wynne, 2004), the failure of entrepreneurial geneticists to define a meaningful unit of analysis of the human genome (Reardon, 2004) and the formation of epistemic communities of climate scientists within an international treaty system (Miller, 2004). In each of these studies, science provides more than rhetorical window dressing for underlying power relations; it shapes the conditions of possibility for the expression of power.

Field Methods and Narrative Analysis

This study takes advantage of the introspection and social mobilization that occurs in the wake of disasters (Oliver-Smith, 2002) to examine interaction within and between two distinct groups: (i) the San Diego Fire Recovery Network (SDFRN) that emerged following the 2003 wildfires near San Diego; and (ii) the established federal, state and county fire agencies. An understanding of the case was informed by meeting summaries, an extensive (500 + messages) email listserv archive of SDFRN communications, planning documents and newspaper articles and editorials. This material was supplemented with interviews, in person and by phone, with key informants associated with these organizations. In-person interviews were recorded and transcribed. Text files of all documents were entered into NVIVO™ qualitative analysis software, which facilitated use of a grounded theory methodology, in which data collection and analysis proceed simultaneously and initial theoretical concepts are modified continuously to reflect and interpret the data (Strauss & Corbin, 1990). Documents were analyzed using a common set of codes, which were then clustered according to whether they came from an SDFRN or agency source.

In another analysis of this case, each group's position on science, management, policy and land use was defined in relation to their conceptions of how nature and society function together in a fire-prone landscape, contrasting two different 'social fire regimes' (Goldstein & Hull, 2007). Using approaches to narrative and discourse analysis (Eckstein, 2003; Roe, 1994), this paper re-examines these positions in relation to their respective visions for sustainability and assembles a composite narrative from the many written and oral accounts told by members of each group. This approach draws on the work of planning analysts who examine planning communications as future and action-orientated narratives that direct attention toward what should be done and who has the authority and legitimacy to act (Sandercock, 2003; Throgmorton, 2003). Planning stories have a problem-solving dimension, focusing on a central inciting event or circumstance and laying out the conflict, crisis and resolution in a way that the characters defined in the story can act upon. Sustainability narratives, like all planning stories, had characteristic scales, spanning a timeline and range of space which had a critical influence on what features become visible and what remain hidden or untold (Soja, 1980). Specific knowledge practices are intrinsic to this story-telling process, as shown in Hajer's (2003) study of how scientific terms such as 'ecological corridor' and 'target types' provide the means and justification to intervene in a nature conceived as 'infrastructure'.

The latter half of this analysis describes the intention and outcome of coalition member efforts to make compelling scientific claims to influence the natural resource agencies. These accounts were drawn principally from observations during three visits conducted to the region, as well as confidential interviews and document analysis. Preliminary drafts of this manuscript were provided to a key informant within each of the respective narrative perspectives and their responses and corrections were incorporated.

The Natural Resource Agencies

The largest of the 2003 wildfires in southern California began on October 25 when a lost hunter set a signal fire in a steep roadless area in rural San Diego County. The conditions were ideal for the outbreak of fire—low humidity, high temperatures and steady high winds, in a landscape already parched by years of drought. County and state firefighters were stretched thin by eleven other fire ignitions in southern California, and this new fire—called the Cedar fire—was difficult to control because it occurred in a highly-flammable shrubland called 'chaparral' that was the dominant vegetation type in San Diego's wildland–urban interface, with its narrow, twisting roads and patchwork of houses. By the next morning the Cedar fire had grown to 100 000 acres—an almost inconceivable spread rate—and began burning into the City of San Diego's suburbs. Local and national media were saturated with dramatic stories and images showing burning homes and landscapes, and area residents demanded that fire agencies explain why the fires could not be controlled. When it was finally extinguished ten days later after the winds died down and rain began to fall, the Cedar fire had become the largest fire recorded in California history at 273 246 acres. Fourteen lives and 2232 homes were lost, and control efforts required 4275 personnel at a cost of $US27 million.

For the next six months after the fire, elected representatives, resource management agencies and firefighting organizations of the region scrambled to respond to continued questioning about whether everything possible had been done to prevent these losses. First, the United States Forest Service and California Department of Forestry (CDF) (2003) released detailed accounts of the 'fire siege' that emphasized the limited resources they had available to deploy against the wildfire. The State of California (2006) and the County of San Diego (2004) then convened formal policy review commissions. Within these documents as well as in public testimony, agency leaders relied on a common story to explain the crisis and identify an appropriate response, which is reconstructed as follows:

> A century of fire suppression and five years of drought were responsible for an unnatural accumulation of dry brush in the San Diego region, creating the potential for an historically unprecedented firestorm. After the Cedar fire began, firefighters lacked the firefighting capacity and surveillance and communications capabilities required to rescue helpless residents whose homes were constructed of highly flammable materials and surrounded by flammable vegetation. To sustain our communities, professional agency land managers should be provided with adequate staffing, enhanced technologies and regulatory powers to reduce risk to life and property by actions such as prescribed burning in the backcountry and creating defensible perimeters around structures.

As noted above, this narrative has strong affinity to the connection between sustainability and security circulating within the field of international relations (Homer-Dixon, 1999; Kaplan, 2000). Centered on the actions of the state agencies, the narrative emphasizes a growing external threat from a natural world that is described instrumentally and mechanistically in terms of fuel accumulation. Providing security requires strengthening border protection between wildlands and vulnerable populations and resources as well as acting pre-emptively beyond this border to reduce fuels through prescribed burning. Only governments have the relevant expertise, so the citizenry should consent to increased regulation and surveillance as well as providing additional tax revenue for equipment, staffing and command and control capacity. Once state powers are augmented in this way, citizens could continue their accustomed lives unmolested by fire, without changing their settlement patterns or land-use practices or assuming any culpability for the crisis.

The rapid and simultaneous expression of this sustainability narrative across a range of local, state and federal agencies had occurred many times before when fires burned homes and aroused the citizenry, such as the 1993 Laguna fire that occurred just up the coast from San Diego. The reappearance of this narrative reflected institutional commitments made at the apogee of European imperialism a century before, when colonial states extended their international reach over natural resources using the rhetorics and practices of forest conservation, irrigation and soil protection (Worster, 1994). Public lands agencies such as the US Forest Service (USFS) were founded in order to conserve valuable timber resources, both from profligate waste by a feckless citizenry and from conflagrations such as the Great Fires of 1910, which killed 85 people as they burned through three million acres in Idaho and Montana (Pyne, 1982). By mid-century the USFS operated a comprehensive system of wildland fire management, funding laboratories in the applied disciplines of forestry, agronomy, hydrology and related agricultural sciences and spreading forestry methods and fire control techniques through co-operative programs with other federal agencies, private firms and the states. The USFS coordinated a national war on wildfire as an off-shoot of the Cold War, as surplus military equipment from World War II and Korea promoted the mechanization of firefighting along with the adoption of military concepts, language and organizational structure (Pyne, 1982).

Since the 1970s, the single-minded pursuit of the war on fire was tempered by recognition that forests that did not burn might accumulate fuels, increasing the risk of uncontrollable wildfire. In order to reduce wildfire risk, resource agencies developed expertise in fuel loading and fire behaviour that allowed them to decide when lightning fires could be allowed to burn or deliberate 'prescribed fires' be set. The top priority of the fire agencies became protection of vulnerable communities at what they termed the "wildlands–urban interface". The Cedar fire was a catalyst for passage of a national law that promised "Healthy Forests" in exchange for funding and permission to aggressively log, burn and thin forests to reduce the accumulation of hazardous fuels. Simultaneously, the response to the Cedar fire was resonant with the concern for territorial security emerging two years after the 9/11 terrorist attacks. Congressional representative Susan Davis reinforced this integration of firefighting with homeland security issues during the California Blue Ribbon Commission hearings examining the southern California fires (Governor's Blue Ribbon Fire Commission, 2004):

I think we must all be very clear that fire fighting in the urban areas, in the wild lands and in the interface is also a homeland security issue. Preparedness must envision the ability to respond to unexpected but massive and even simultaneous events in the future.

SDFRN

While the Cedar fire was still burning, email went out to mobilize San Diego's environmental activists, an engaged and intricately networked community that had spent a decade conducting advocacy and planning to protect open space and conserve the region's many endangered species. Eighty of these conservation activists, land managers and biological consultants gathered at a hastily assembled meeting on 30 October 2003, where they agreed to take part in an association that they named the San Diego Fire Recovery Network (SDFRN). For the next four months SDFRNers remained in nearly daily contact with one another, defining their collective perspective on the causes and consequences of the fire into a narrative that was radically different than the agencies, reconstructed as follows:

> *The chaparral ecosystem is dynamic and self-regulating, and the Cedar fire was a normal, natural event, an inevitable and recurring feature within an ecosystem that has evolved with fire over millennia and needs large, stand-replacing fires. Homes sprawled throughout the backcountry only added extra fuel to the fire. While fire frequency has varied since human arrival in the region, humans have never been able to control or prevent chaparral fires, and efforts to reduce fire risks through controlled burning, clearing, or re-vegetation have only caused conversion of this vulnerable, globally significant biodiverse chaparral into highly fire-prone non-native grassland. The people of the region should collectively mobilize to perform restoration efforts that emphasize native species and pre-settlement conditions adapted to fire, as well as to catalyze land use planning that prevents sprawl.*

This fire narrative has strong affinity with Dalby's (2002) previously noted approach to coupling sustainability and security. The Cedar fire is placed within an evolutionary context spanning from before human occupation to the indefinite future, a time span that is inclusive of human occupation but not exclusive to it. Over this longer time span, big fires are normal and natural occurrences that serve to maintain biodiversity, measured in relation to global ecology rather than anthropocentric worth. A precautionary approach to manipulating environmental conditions is advised since natural systems are dynamic, self-regulating and only partly understood, and efforts to control them may cause ecological degradation. Rather than conforming with the agency narrative's spatial imaginary of vigilance at the border separating people and their resources from external threats, the SDFRN narrative suggests that sustainable human communities exist within healthy natural ecosystems. Since destructive wildfires are the inevitable result of living out of ecological balance and will occur regardless of governmental fire fighting capacity or fuel accumulation, citizens have little choice but to bring their land-use practices into harmony with fire's dynamic rhythms or continue to pay a steep price in property and lives. This adaptation is the primary responsibility of civil society, with only a supporting role for governmental coordination of collectively agreed-upon constraints on land use.

In composing this narrative, SDFRNers defined the Cedar fire within the context of evolutionary time and the patterns of global biodiversity, and interpreted the landscape as a self-regulating, dynamic ecosystem. Their reliance on ecological science was accompanied by frequent reference to their own field observations of the distribution of local flora and fauna, acquired through years of patient observation around San Diego county. For example, many SDFRNers were part of San Diego's active community of naturalists, whose dedication was shown in the publication of a 645-page *Bird Atlas of San Diego County* (Unitt, 2005), produced by 400 volunteers who spent over 55 000 hours conducting field observations between 1997 and 2002.[3] This capacity to join together ecology and natural history facilitated conversation and forging of common purpose across a group composed of environmental educators and activists, naturalists, ecologically trained land managers and consultants, and research ecologists.

Table 1 compares features of the contrasting narratives of government agencies and SDFRN in the aftermath of the Cedar fire. Structured along similar thematic lines, the narratives corresponded to radically different conceptions of sustainability.

SDFRNers concurred that the Cedar fire provided an opportunity for them to demonstrate that public safety required reorientating settlement patterns and land-use practices to accommodate periodic and inevitable fires. They were also motivated by a collective sense of urgency to develop alternatives to state-sponsored fire remediation efforts that government agencies were proposing after the wildfire. Not only would erosion-control treatments and prescribed burning distract the public from what was required to achieve sustainability, these efforts would catalyze the arrival of additional backcountry homeowners who would demand that agencies burn and thin, increasing the disruption of ecological systems in futile attempts to alter the timeless return of huge chaparral fires.

Given this concern, SDFRNers decided to focus on providing the natural resource agencies with scientific advice, reasoning that the basis of their own political influence and credibility was their knowledge of ecological science and the

Table 1. Contrasting narratives of government agencies and SDFRN in the aftermath of the Cedar fire

Features of narrative	Agency sustainability narrative	SDFRN sustainability narrative
Temporality	From the origins of the agencies to resolution of fire problem in the immediate future	From the evolution of species into the indefinite future
Spatiality	Division or boundary maintained between area with excess fuels and human community	Integration of human communities within natural landscape
Cause of security threat	External resource imbalance (excess of fuels)	Human actions within ecological systems
Knowledge and control	Environment is well understood and humans are capable of manipulating it to their benefit	Partial knowledge of ecological dynamics require precaution
Leadership and governance	Government agencies protect people and resources from fire	Civil society takes lead in maintaining healthy relationship between human communities, ecological systems and fire

area's natural history. Their hope was that once the agencies understood the poor scientific basis for erosion control and prescribed burning, the agencies would redirect resources toward activities compatible with SDFRN's sustainability narrative, such as protecting sensitive ecological sites and allowing native chaparral ecosystem to naturally regenerate over time. As the following account shows, things did not work out this way—the agencies rejected SDFRN's arguments, concluding that the scientific advice provided by the community group was neither legitimate nor credible, and that the policy alternatives SDFRNers proposed were a distraction from the need to perform critical and time-sensitive landscape interventions.

Scientific Advice on Erosion Control Measures

Destructive landslides are as regular a feature of disaster coverage as catastrophic fires in the newspapers of southern California, where expensive houses cling to steep mountain sides in a tectonically active landscape. Since most of the year's rainfall comes during spring rains that follow the fire season, concern about landslides arose immediately after the Cedar fire. Government agencies were quick to respond to this heightened concern about erosion by proposing to hire firms that would broadcast seeds on the landscape and 'hydromulch', which involves spraying a bright green papier mâché-like substance over burned slopes. SDFRNers were alarmed by these proposals, reasoning that this would interfere with chaparral's evolutionary capacity to recolonize burnt areas and facilitate the irreversible establishment of highly flammable non-native grasslands in chaparral's place. In addition, they were concerned that these highly visible remediation projects would reassure residents that they could rely on government agencies to protect them from their environment, recent experience during the fires notwithstanding. As one SDFRNer put it, erosion control measures:

> ... tend to give people a false sense of security that something has been done to reduce the risk of erosion and slope failures, and tend to perpetuate the myth that human intelligence supersedes the collective intelligence of over 2 billion years of evolution on Earth.

SDFRNers initiated a dialogue with one erosion control company, and a representative of this firm attended an SDFRN meeting and provided a packet of journal articles that demonstrated that hydromulching and planting quick-growing vegetation stabilizes the soil surface and reduces erosion immediately after a fire. SDFRNer's responded skeptically—one wrote on the group's listserv that these articles were "industry-generated and financed", "biased to the point of deception" and "little more than sales brochures in academic or pseudo-academic clothing". Recognizing a distinction between his own scientific practices and those of the erosion control consultants, this SDFRNer also noted that "... the 'journals' are oriented to traditional concepts and applications more than questioning conventional practice".

SDFRNers decided to attempt to change agency erosion control practices by providing scientific advice to the Burned Area Emergency Response (BAER) team, an interagency group of fire rehabilitation specialists who were flown in to San Diego to prioritize all federally funded erosion control activities. SDFRNers were concerned that the BAER team's hydrologically orientated protocol and rushed timetable would lead to heavy-handed landscape modification, ignoring the habitat requirements of vulnerable species and interfering with chaparral's capacity to recover on its own. To help the BAER team appreciate the sensitivity of regional flora and fauna, ten SDFRNers worked furiously for two weeks to compile their decades of local field experience into a 36-page guide to the location and condition of vulnerable species and habitats (San Diego County Biological Resource Researchers, 2003). The guide urged the BAER team to take a precautionary approach to habitat alteration and to initiate intensive long-term ecological monitoring—as one SDFRNer concluded; "This is a huge ecological experiment that must be carefully monitored over both long and short terms so we can learn something". SDFRNers agreed that the best member of their group to deliver the report to the BAER team was the Forest Supervisor of the Cleveland National Forest, where much of the Cedar fire had burned. The Forest Supervisor had played a critical role in forming SDFRN, an action that was characteristic of her unorthodox collaborative management style and commitment to biodiversity conservation.

When the Forest Supervisor attempted to hand SDFRN's species and habitat guide to the leader of the BAER team, he refused to include it as an appendix to the BAER team's official report, responding that BAER teams worked autonomously and did not accept public comment that would delay their efforts and compromise the professional integrity of their recommendations. Astounded by what she later described as the BAER team leader's inflexibility and unwillingness to adapt to local conditions, the Forest Supervisor argued that he should accept the guide, but the answer was final—and the BAER team leader carried the issue further by writing an administrative complaint against the Supervisor for attempting to force the guide on him. Within weeks, the Pacific Southwest regional forester involuntarily transferred the Forest Supervisor to an administrative position in northern California. Rather than move, the Forest Supervisor retired. She continued to work closely with SDFRN—indeed, she was able to devote more of her time to the effort—but association with this controversy and the loss of ready access to the staff and resources of the Cleveland National Forest cost the group dearly over the months to come. SDFRN's compilation of the location and vulnerability of species and habitats was never used by the BAER team, which filed their recommendations for slope stabilization and hazardous tree removal and then departed the region.

Scientific Advice on Prescribed Fire

SDFRNers agreed that large-scale prescribed burning in chaparral caused ecological harm without providing any public safety benefits. This was an unusual position for a group of environmentalists and ecologists to take, since prescribed burning had long been heralded as an effective means to address large destructive fires that were the consequence of a century of fire exclusion on forested public lands across the country (Busenberg, 2004). Yet some commentators have questioned this reformist story that fire should be put back on every landscape. As Pyne (2004, p. 11) put it, this "absolutism ... is simplistic in ways that make reform more difficult and that, having become canonical, it tends to exclude all the other stories". From the beginning of their engagement with fire, SDFRNers attempted to identify the appropriate fire policy by identifying the appropriate "fire regime" for the area, which they understood to mean the characteristic frequency, season, severity and size of fires on a landscape, which are driven by

climate and biophysical setting (e.g. vegetation, topography and soils). The challenge they faced was that chaparral fire regime science had long been riven by a
disagreement among the field's two leading researchers. Richard Minnich of the
University of California Riverside (1983) had adopted the reformist interpretation
of fire-starved ecosystems, arguing that creating a vegetative patchwork through
prescribed burning would reduce wildfire risk while restoring southern
California's chaparral to health after a century of fire suppression. Minnich had
been challenged by Jon Keeley (Keeley et al., 1999) of the US Geological Service,
who asserted that fire suppression has not altered the frequency of catastrophic
fires over the last century, since unstoppable fires burn through chaparral of
any age if moisture was low and winds were high. Years of high-profile debate
had only resulted in the hardening of positions on both sides and a widely
known personal animus between the two scientists, who regularly traded accusa-
tions of misrepresentation and bias.

Since both Minnich and Keeley were respected scientists, SDFRNers sought
to consider the merits of both sides of the controversy by organizing a scientific
forum to evaluate the two positions and by asking both scientists to respond to
queries on the SDFRN listserv. After a few weeks of deliberation SDFRNers concluded that a critical distinction between the two scientists was that Minnich, a
biogeographer by training, analyzed chaparral simply as a fuel source, rather
than as a diverse ecological community. In contrast, Keeley, who was an ecologist
by training, emphasized the possible ill-effects of too frequent burning of chaparral, which could lead to a "type-conversion" of chaparral to non-native grasslands
of little value to native wildlife. SDFRNers also were concerned that Minnich sanc-
tioned burning that could potentially lead to irreversible alteration of native chaparral into non-native grasslands. As one SDFRNer concluded: "We should use
restraint in our desires to 'do something' and always err on the side of caution
when making recommendations on how best to manage these diverse, complex,
and unpredictable ecosystems".

By early 2004 SDFRN's emerging narrative incorporated Keeley's position
that unstoppable wind-driven wildfires were inevitable in chaparral regardless
of fuel accumulation, a position that left San Diegans no choice but to bring
land-use practices into harmony with fire's dynamic rhythms or continue to
pay a steep price in property and lives. Accordingly, SDFRNers became concerned
when a county land manager who had managed fires for decades began advocating prescribed burning to a variety of influential audiences, including San Diego
County's elected supervisors. For SDFRNers, the county land manager's efforts
appeared nakedly self-serving—as one SDFRNer put it, "Prescribed burning is
a job that gets funding that buys equipment that pays salaries". If his views
remained unchallenged, SDFRNers agreed that he could undermine their whole
initiative. As one SDFRNer put it: "Preaching that firestorms are preventable if
only the government would chip and burn our precious natural resources provides false hope and sets the stage for future disasters".

Once again, SDFRNers attempted to influence agency policy by providing
scientific advice, this time by organizing a scientific peer review of the "Wildland
Task Force Report", a document written by the county land manager that the
county had released in August 2003, a few months before the Cedar fire. This
report had cited Minnich's research to conclude that prescribed fire was required
to redress the unnatural accumulation of woody biomass. An SDFRNer invited
Minnich's antagonist Keeley and three of his colleagues to comment, who

obliged by providing critiques that accused the report of ignoring Keeley's oft-published alternative to Minnich's view, misrepresenting evidence addressing whether fires were controlled by fuel load or wind conditions, and fabricating bibliographic citations in order to support a preference for prescribed burning. SDFRNers attached these critiques to a hard-hitting cover letter and press release that concluded that task force report was "woefully inadequate", "biased in its treatment of available scientific information" and "flawed in many of its assumptions". In place of what they regarded as a scientific travesty, SDFRN suggested that county should formally withdraw the task force report and adopt an approach more in accord with SDFRN's approach to sustainability—as they concluded: "The new report should address, based on the best available information, the most effective, cost-efficient, and sustainable approaches for reducing risks to human life and property at the wildland–urban interface".

By early February 2004, the letter was in the hands of the San Diego County Chief Administrative Officer, county supervisors and the media. Publicly, the county's response to SDFRN's peer review was mild. Both the county land manager and a county supervisor told a reporter from the *San Diego Tribune* (Balint, 2004) that while there may have been some minor errors in the county's report, these errors did not justify withdrawing the report altogether. However, behind the scenes, infuriated administrators at county resource agencies directed their employees to discontinue attending SDFRN meetings or participating in SDFRN workshops. Word of this boycott spread within San Diego's environmental community, discouraging potential sponsors of SDFRN activities. Alarmed SDFRNers secured a meeting with top administrators within the County Department of Planning and Land Use in early May 2004, but found that county administrators did not share their agenda of reconciliation. The administrators accused the SDFRNers of "tailgunning" efforts to address urgent public safety priorities, and threatened that no one associated with SDFRN would ever get monitoring or research contracts from the county again, a threat that was particularly troublesome to SDFRNers whose consulting livelihoods depended on government contracts.

SDFRNers vowed to stand strong after the meeting with the county administrators, but key members of SDFRN began to withdraw from participation in the group. An attempt was made to rally the remaining SDFRNers and reorientate the group as a forum for all perspectives on fire within the county. While a flurry of workshops and speaking engagements were scheduled over the next six months, activity dropped off significantly on the listserv and at meetings, and SDFRN became a contact list for occasional mass emails. By the first anniversary of the Cedar fire in October 2004, with memory of the fires fading in the county, the principal legacy from the Cedar fire was that firefighting capacity in the San Diego region was better co-ordinated and equipped, government-directed tree and brush removal projects were implemented, and a range of new regulatory tools were available to increase the security periphery between landowners and flammable vegetation (Gross, 2005). Life continued largely as it had before in San Diego county, despite the warning from SDFRN of the threat posed by the fire next time.

Narrative Incommensurability

San Diego's natural resource agencies were more than merely unconvinced by SDFRN's science-based arguments—they described coalition claims as irrelevant and even deliberately inflammatory. The depth of this divide between SDFRN and the natural resource agencies was most apparent during SDFRN's peer review of the county's Wildland Task Force Report. SDFRN advocated precaution, since complex ecological interactions could never be completely understood or predicted. For county land managers and the other resource agencies, precaution meant unacceptable inaction, and SDFRN's recruitment of a group of ecologists to declare that the county's most experienced fire manager was scientifically unfit was especially unwelcome at a time when the agencies were mobilized to restore public trust in their own capacity to restore order and maintain control.

The resource agencies' instrumental knowledge practices were situated in a century-long institutional objective of imposing machine-like predictability on the nation's forests by managing fuel levels and reducing hazard. This sustainable resource management narrative was temporally and spatially disjunctive with SDFRN's narrative, in which humans had to accommodate themselves to complex and unpredictable fire dynamics that played out over evolutionary time. For SDFRNers, society had to learn to accommodate fire or suffer the consequences, because humans would never be able to control large fires in a landscape that was both highly dynamic and vulnerable to unanticipated and undesired change. The agencies' singular focus on human safety and their apparent disregard for the ecological effects of hasty remediation activities were both scientifically and morally unacceptable to SDFRNers, within their conception of seeking sustainability within an encompassing ecology, and the need for civic expertise and broader participation in decision making.

The dismissal of SDFRN's species and habitat guide by the Burned Area Emergency Response team highlights how differences over what was appropriate knowledge to achieve sustainability extended to include differences over who could claim to be a legitimate and credible expert. SDFRNers had assumed that the BAER team would welcome a compilation of their field observation of habitats and species acquired through years of patient observation around San Diego county. However, the long-established methodology of the BAER team was to assemble a team of experts drawn from the applied sciences of hydrology and soil science, experts whose objectivity was ensured by their association with the agencies rather than their personal experience and commitment to a specific place. For the BAER team leader, SDFRN's amalgam of ecological and local knowledge was not credible science. Instead, it was unsolicited public comment, an unwelcome intrusion into an activity whose scope was officially limited to ensuring public safety after an emergency through erosion control and other interventions. As the BAER team leader protested, taking this guide into account would only delay the BAER team's effort and compromise their professional integrity. This commitment to the exclusive legitimacy of agency scientists was coupled with a commitment to the exclusive authority of the state, in contrast to the paramount role of civil society expressed in SDFRN's sustainability narrative.

SDFRN's scientific advice fell flat when pitched over this discursive and epistemic divide. Furthermore, the controversy that the community coalition engendered through their scientific intervention led to the dissolution of their initiative, and even threatened SDFRNer careers and livelihoods. This discouraging outcome to SDFRN's attempt to steer the agencies toward their approach to sustainability underscores that just as local and traditional communities have their own culturally situated knowledge practices (Wynne, 1996), government agencies also have long-established commitments to specific configurations of knowledge that are grounded in their own organizational history and institutional relationships. Even though the legitimacy of these agencies rests in part upon formally credentialed scientific knowledge and they labor under a legal obligation to use the 'best available science', this case suggests that agencies may resist the compelling force of a scientific argument when it is incongruent with their organizational knowledge practices and discursive frameworks. Just as a synthesis of knowledge, purpose and meaning is fundamental to agency identity, the scientific claims that underpin an alternative social order are more than merely meaningless—they are threatening. In this sense, people and ideas that cannot be aligned within an existing narrative are more than incompatible, they are "enemies" to the integrity of this specific arrangement of society and nature (Latour, 2004).

For both SDFRN and the natural resource agencies, their particular knowledge practices did much more than provide rhetorical window dressing for their pursuit of power (e.g. Flyvbjerg, 1998)—these knowledge practices provided a way to imagine what the future could hold, while delimiting the limits of possibility of citizenship and governance in which ideas such as sustainability had meaning and significance. Recognizing how the ambivalence of sustainability narratives is sustained by epistemic difference does not preclude the possibility of questioning a dominant narrative while still harvesting a 'security bonus', although the appropriate strategies for this kind of intervention have barely been explored within co-production research, beyond an acknowledgement for the need for "civic epistemology" (Jasanoff, 2005; Miller, 2005). Long-term studies suggest that discursive transformation can occur, although adoption of new narratives must be accompanied by shifts in knowledge practice and social identity. For example, Agrawal (2005) described how villagers in rural India were discursively reconstructed from greedy, ignorant peasants who were an obstacle to rational central planning of forest resources into resourceful practitioners of local knowledge, a shift that made decentralized, community-based forest management possible. Agrawal's success story and the failed effort in San Diego both underscore how the incommensurability of different forms of knowledge is reinforced by differences in individual subjectivity and institutional relations.

Conclusion

The scientific arguments that SDFRN deployed failed to convince San Diego's natural resource agencies to discard a century-old conception of fire security and sustainability in favor of an ecological alternative. SDFRNers agreed that this outcome was the unfortunate consequence of dealing with prickly bureaucratic personalities who had the power to ignore inconvenient truths. Their interpretation, which was consistent with the commonly held idea that scientific evidence functions as a *post-hoc* rationale for pre-existing policy, affirmed group solidarity during a stressful time. However, a more symmetrical interpretation emerges after considering the way that knowledge practices are vested within sustainability narratives. For both SDFRN and the agencies, distinct knowledge practices underpinned distinct conceptions of sustainability, and were an outcome of their efforts to bring those conceptions into being. The absence of jointly accepted scientific practice and expertise not only made it impossible for the two sides to resolve the ambivalence between their different conceptions of security and sustainability, it reinforced these divisions, as each side perceived the seemingly irrational claims of the other side as an obstacle to achieving sustainability. In this way, the incommensurability of knowledge practices maintains the ambivalence of sustainability.

Notes

1. Architect Sim Vander Ryn, quoted in Dowie (1995, p. 205).
2. 'Progressive era' describes a period of US governmental reform from the 1890s to the 1920s, typified by a utilitarian approach to natural resources that provided for "the greatest good for the greatest number for the longest time", a phrase coined by Gifford Pinchot, the first Director of the US Forest Service.
3. Figure cited by publisher on http://www.sdnhm.org/research/birdatlas/ (accessed 8 April 2005). This amounts to an average of seventeen full eight-hour days per individual, or having an individual in the field watching birds around the clock over that entire time period.

References

Agrawal, Arun (2005) *Environmentality: Technologies of Government and the Making Of Subjects* (Durham, NC: Duke University Press).
Balint, Kathryn (2004) Scientists say their work has been misrepresented, *The San Diego Union-Tribune*, 13 February.
Barnett, Jon (2001) *The Meaning of Environmental Security* (New York and London: Zed Books).
Busenberg, George (2004) Wildfire management in the United States: the evolution of a policy failure, *Review of Policy Research*, 21(2), pp. 145–156.
Coburn, Jason (2003) Bringing local knowledge into environmental decision making, *Journal of Planning Education and Research*, 23(4), pp. 420–433.
Dalby, Simon (2002) *Environmental Security* (MN: University of Minnesota Press).
Dowie, Mark (1995) *Losing Ground: American Environmentalism at the Close of the Twentieth Century* (Cambridge, MA: MIT Press).
Eckstein, Barbara (2003) Making space: Stories In The Practice Of Planning Story and Sustainability, in: Barbara Eckstein & James Throgmorton (Eds) *Story and Sustainability: Planning, Practice and Possibility for American Cities*, pp. 13–36 (Cambridge, MA: MIT Press).
Ellul, Jacques (1964) *The Technological Society* (New York: Knopf).
Flyvbjerg, Bent (1998) *Rationality and Power: Democracy in Progress* (Chicago, IL: University of Chicago Press).
Foucault, Michel (1977) *Discipline and Punish: The Birth of the Prison* (trans. by Alan Sheridan) (New York: Vintage Books).
Fuller, Steve (2000) *Thomas Kuhn: A Philosophical History For Our Times* (Chicago: University of Chicago Press).
Goldstein, Bruce E. & Hull, R. Bruce (2007) Socially explicit fire regimes, *Society and Natural Resources*, 00, pp. 00–01, in press.
Governor's Blue Ribbon Fire Commission (2004) *Report to the Governor* (Sacramento, CA: State of California).
Gross, George Alan (2005) Policies, strategies evolving, *San Diego Union Tribune*, 25 October.
Hajer, Maarten (2003) A frame in the fields; Policymaking and reinvention of politics, in: Hajer, Maarten & Wagenaar, Hendrik (Eds) (2003) *Deliberative Policy Analysis: Understanding Governance in the Network Society* (Cambridge, UK: Cambridge University Press).
Haraway, Donna J. (1996) *Simians, cyborgs and women: the reinvention of nature*, (London: Routledge).
Hess, David J. (1997) *Science studies: an advanced introduction* (NY: University Press).
Homer-Dixon, T. (1999) *Environment, Scarcity, and Violence* (Princeton, NJ: Princeton University Press).
Jasanoff, Sheila (1987) Contested boundaries in policy-relevant science, *Social Studies of Science*, 17, pp. 195–230.
Jasanoff, Sheila (2004) Ordering knowledge, ordering society, in: Sheila Jasanoff (Ed.) *States of Knowledge: The Co-Production of Science and Social Order*, Chapter 2 (London: Routledge).
Jasanoff, Sheila (2005) *Designs On Nature: Science and Democracy in Europe and the United States* (Princeton, NJ: Princeton University Press).
Kaplan, R. D. (2000) *The Coming Anarchy: Shattering the Dreams of the Post Cold War World* (New York: Random House).

Keeley, J. E., Fotheringham, C. J. & Morais, M. (1999) Reexamining fire suppression impacts on brushland fire regimes, *Science*, 284, pp. 1829–1832.

Kingdon, John W. (1984) *Agendas, alternatives, and public policies* (Boston, MA: Little, Brown and Co.).

Kuhn, Thomas S. (1970) *The structure of scientific revolutions*, 2nd edn, enlarged (Chicago, IL: University of Chicago Press).

Latour, Bruno (2004) *Politics of Nature: How to Bring the Sciences Into Democracy* (Cambridge, MA: Harvard University Press).

Majone, G. (1989) *Evidence, argument and persuasion in the policy process* (New Haven, CT: Yale University Press).

Miller, Clark (2004) Climate Science and the Making of Global Political Order, in: Sheila Jasanoff (Ed.) *States of Knowledge: The Co-Production of Science and Social Order*, pp. 46–66 (London: Routledge).

Miller, Clark (2005) New civic epistemologies of quantification: Making sense of indicators of local and global sustainability, *Science, Technology and Human Values*, 30(3), pp. 403–432.

Minnich, Richard A. (1983) Fire mosaics in southern California and northern Baja California, *Science*, 219, pp. 1287–1294.

Newton, Julianne L. & Freyfogle, Eric T. (2005) Sustainability: A dissent, *Conservation Biology*, 19(1), pp. 23–32.

O'Riordan, T. (1988) The Politics of Sustainability, in: R. K. Turner (Ed.) *Sustainable Environmental Management: Principles and Practice*, pp. 29–50 (London: Belhaven Press).

Oliver-Smith, Oliver (2002) The brotherhood of pain: Theoretical and applied perspectives on post-disaster solidarity, in: Susanna M. Hoffman & Anthony Oliver-Smith(Eds) *Catastrophe and Culture: The Anthropology of Disaster*, pp. 156–172 (Santa Fe: NM School of American Research Press).

Popper, Karl R. (1959) *The Logic of Scientific Discovery* (London: Hutchinson).

Prins, G. (2004) AIDS and Global Security, *International Affairs*, 80(5), pp. 931–952.

Pyne, Stephen J. (1982) Fire in America: *A Cultural History of Wildland and Rural Wildlife* (Princeton, NJ: Princeton University Press).

Reardon, Jenny (2004) *Race To the Finish: Identity and Governance in An Age of Genomics* (Princeton, NJ: Princeton University Press).

Redclift, Michael R. (2006) Sustainable Development (1987–2005): An oxymoron comes of age, *Horizontes Antropológicos*, 12(25), pp. 65–84.

Roe, E. (1994) *Narrative policy analysis: theory and practice* (Durham, NC: Duke University Press).

San Diego County Biological Resource Researchers (2003) *A summary of affected flora and fauna in the San Diego County fires of 2003* (San Diego: San Diego County Fire Recovery network).

San Diego Regional Emergency Preparedness Task Force (2006) *San Diego Regional Emergency Preparedness Task Force Final Report*, San Diego, CA: San Diego Country. 10pp. Available at: http://www.sandiego.gov/fireandems/ accessed 10/2/07.

Sandercock, Leonie (2003) Out of the closet: The importance of stories and storytelling in planning practice, *Planning Theory and Practice*, 4(1), pp. 11–28.

Scott, James C. (1998) *Seeing Like a State: Why Certain Schemes to improve the Human Condition have Failed* (New Haven, CT: Yale University Press).

Shapin, Steven & Schaffer, Simon (1985) *Leviathan and the air pump: Hobbes, Boyle and the experimental life* (Princeton: Princeton University Press).

Soja, Edward (1980) The socio-spatial dialectic, *Annals of the Association of American Geographers*, 70, pp. 207–225.

Strauss, A. & Corbin, J. (1990) *Basics of Qualitative Research; Grounded Theory Procedures and Techniques* (Newbury Park: Sage).

Throgmorton, James A. (2003) Planning as persuasive storytelling in a global-scale web of relationships, *Planning Theory*, 2(2), pp. 125–151.

United States Forest Service and California Department of Forestry (2003) *The 2003 San Diego County Fire Siege Fire Safety Review*, United States Forest Service and California Department of Forestry. Available at http://www.fs.fed.us/r5/cleveland/documents/documents/sandiegocountyfinal.pdf (10/2/07).

Unitt, Philip (2005) *San Diego County Bird Atlas: Proceedings of the San Diego Natural History Museum Number 39* (Temucula, CA: Ibis Publishing).

Voß, Jan-Peter, Newig, Jens, Kastens, Britta, Monstadt, Jochen & Nölting, Benjamin (2007) Steering for Sustainable Development: A Typology of Problems and Strategies with Respect to Ambivalence, Uncertainty and Distributed Power, *Journal of Environmental Policy and Planning*, 9(3/4), pp. 227–244.

Walker, Gordon & Shove, Elizabeth (2007) Ambivalence, Sustainability and the Governance of Socio-Technical Transitions, *Journal of Environmental Policy and Planning*, 9(3/4), pp. 227–244.

Waterton, Claire & Wynne, Brian (2004) Knowledge and political order in the European Environment Agency. in: Sheila Jasanoff (Ed.) *States of Knowledge: The Co-Production of Science and Social Order*, pp. 87–108 (London: Routledge).

Williams, Colin C. & Millington, Andrew C. (2004) The Geographical Journal, *The Diverse and Contested Meanings of Sustainable Development*, 170(2), pp. 99–104.

Worster, Daniel (1994) *Nature's Economy: A History of Ecological ideas* (Cambridge, UK: Cambridge University Press).

Wynne, Brian (1996) May the sheep safely graze? A reflexive view of the expert-lay knowledge divide, in: Scott Lash, Bronislaw Szerszynski & Brian Wynne (Eds) *Risk, environmental and modernity: towards a new ecology*, pp. 44–83 (London: Sage Publications).

Working Towards Sustainable Development in the Face of Uncertainty and Incomplete Knowledge

Introduction and Overview

The vision of sustainable development must, by definition, include both long-term considerations and the global dimension (Grunwald & Kopfmüller, 2006; WCED, 1987). Pursuing this vision implies that societal processes and structures should be re-orientated so as to ensure that the needs of future generations are taken into account and to enable current generations in the Southern and Northern

Hemispheres to develop in a manner that observes the issues of equity and participation. Since a feature inherent in the Leitbild of sustainable development is consideration of strategies for shaping current and future society according to its normative content, *guidance* is necessary and the ultimate aim of sustainability analyses, reflections, deliberations and assessments. The latter should result, in the last consequence, in *knowledge for action*, and this knowledge should motivate, empower and support 'real' action (von Schomberg, 2002).

The first task of this paper is to analyse the gap between the prerequisites for this knowledge being made usable for working towards sustainable development, on the one hand, and the possibilities for satisfying these requirements on the other. Its second task is to reflect on the consequences and impact of this gap for guidance strategies. The issue of the *uncertainty of knowledge* is at the heart of this gap. A specific aspect of this uncertainty consists of the *ambivalence* of the basis of sustainability assessments. In many cases there is no sharp borderline between facts and values (e.g. Jasanoff, 2004), with the consequence that epistemic and normative statements and questions cannot be separated strictly (Nowotny *et al.*, 2001). This situation has serious implications and consequences for steering towards sustainability, especially in cases where societal conflicts, diverging judgements and contested futures (Brown *et al.*, 2000) are involved. This is frequently the case in sustainability policies. This situation supports the necessity to look at these issues in more analytical detail.

The main parts of the analysis given below and the implications of the arguments used may be summarized as follows:

- Uncertainties from different sources merge in the context of sustainable development, dramatically increasing the importance and relevance of this issue. Examples are the long-term nature of sustainable development, its close ties to both nature and society, and its integrative function among incoherent subtopics.
- Guiding societal transformation in the direction of sustainable development, therefore, is not feasible within a classical planning approach; new concepts for change management and approaching the future are needed.
- There is no reason merely to lament about uncertainties, especially not those concerning our knowledge of the future, because uncertainties represent the other side of the coin to the openness of the future and of our options for shaping it.
- Uncertainties, however, dramatically change the nature of which strategies for guiding change are adequate. They both require and permit learning over time, just as they both require and permit measures to be adapted and modified, for example, as a result of the monitoring of processes.
- A prerequisite for exploiting the chances offered by this situation is to have specific implementation strategies accompanying the measures adopted to promote sustainable development, in particular, strategies that foresee adaptation and 'online' modification by monitoring the effects of the initial measures.

On the one hand, distinction will be made between uncertainty and risk. While risk is a quantifiable parameter where there is both significant scientific knowledge about the probability of the occurrence of certain effects and reliable knowledge about the nature and the extent of the possible harm, uncertainty is characterized by a limited quantifiability, a lack of knowledge, epistemic uncertainty and/or unresolved scientific controversies (von Schomberg, 2005). For the purposes of this paper, however, uncertainty must be distinguished also from mere speculation. There should be some 'evidence' supporting the assumptions being considered, but not enough to prove them scientifically.

The analysis given in this paper is conceptual. Different lines of thought from humanities and sustainability research are put together, and a novel, structured framework is offered for further work in the field of sustainability governance facing non-removable uncertainties. The literature taken into account includes relevant debates in social and political sciences, in planning theory, in the STS world, and in sustainability science (Kates *et al.*, 2000), aiming at analysing the consequences and implications for sustainability governance in the situation of severe uncertainties. The function of the paper in this collection is to provide an overview of uncertainty issues that can become virulent in governance for sustainable development.

Guiding change to achieve sustainable development requires the availability of knowledge about different types of research and reflection. The concept of shaping technology in connection with sustainable development will be used to identify and clarify issues of uncertainty and incompleteness of knowledge. Cross-cutting remarks to this issue will be supplemented by a conceptual analysis of different types of uncertainties, the reasons for them and their degrees. The final section will discuss the consequences of this situation and the options for dealing with them constructively in guiding processes.[1]

Working Towards Sustainable Development and the Need for Knowledge

Sustainable development is a normative societal principle rather than a descriptive scientific concept. None the less, science and empirical research are indispensable in order to identify weak spots in sustainability, to elaborate strategies for achieving sustainable development and for monitoring their impacts. All scientific research carried out in connection with the identification, appraisal and development of measures to overcome sustainability weak spots belongs in this category. The knowledge required to analyse and assess sustainability and to provide guidance is available at quite different levels and in various manners. There are several patterns of knowledge production that contribute to sustainable development and that are related to specific activities in science and research (Grunwald, 2004).[2]

- *Empirical observation of developments over time*: The temporal dimension of sustainability implies that problems of sustainability can only be ascertained by the observation, over time, of certain parameters (such as the chemical composition of the atmosphere, of soils or of water) or of societal phenomena (such as the development of education). Science is therefore in demand to provide the means for empirical observation of pertinent indicators over long periods of time.
- *Recognition of cause-and-effect relationships*: Knowledge of cause-*and*-effect relationships forms the basis for models, yet it can also guide our actions in dealing with weak spots in sustainability and help identify the appropriate measures for political intervention. The study of causal relations extends to basic research, such as atmospheric chemistry or political science. At the level of complex systems, such as in nature, in society and at the interface between

them, research on causal relations is particularly difficult, especially because of circular feedback and the dynamics of self-organization (Schellnhuber, 1999).

- *Providing knowledge-based, coherent pictures of the future*: In many cases, illustrations of future developments, scenarios, or if possible, predictions are necessary in order to assess whether there will be sustainability problems (examples are the development of the global climate and of the world population). Although it is impossible to experiment with the real world, predictions can be made by modelling and simulating the system under consideration (Kates *et al.*, 2000; Rotmans, 1999; Schellnhuber, 1999). In many fields, however, future outcomes cannot be predicted because either human actions, decisions and behaviour are involved to a large extent, or simply the data are imperfect or cause-*and*-effect relationships have not been determined. In these cases, scenarios as illustrations of *possible* futures are used in order to structure the spectrum of further developments, to identify 'worst' and 'best' cases, to create common visions, and to acquire knowledge to draw up action strategies.
- *Assessing and judging sustainability*: It cannot be ascertained from empirical observation or simulation alone whether certain effects and tendencies should be interpreted as sustainability problems. Criteria have to be defined, according to which observations can be classified, as more or less relevant for sustainability or even as sustainability weak spots. The problem of judgement extends to ethical questions of responsibility for the future and to the theory of justice (Brown-Weiss, 1989; Sen, 1987). An essential challenge is the operationalization of such 'grand' theories, which have to be made operable at the level of concrete and measurable indicators. The normative dimension of making assessments and judgements, however, restricts the applicability of familiar quantitative scientific approaches such as multi-criteria decision analysis (MCDA) and leads to the necessity of discursive elements of assessment.
- *Ranking of sustainability weak spots*: The analysis of the sustainability situation also has to concern itself with assessing the *urgency* of certain sustainability problems compared to others, and with setting well-founded priorities. Controversy, however, does surround the possibilities of making scientifically well-founded statements in this field and the limitations on them (see also the point above concerning the limitations on quantitative scientific approaches).
- *Interpretation of the results*: The results of modelling, simulations and assessments must be interpreted carefully. Models and simulations are based on specific presuppositions, for example, with regard to the definition of system boundaries, the cause-*and*-effect relationships taken into account or neglected, or assumptions about the interface between a system and its environment. In addition to the possibilities for clarification immanent in the respective disciplines, hermeneutics and the philosophy of science are also needed to create the 'meta-knowledge' about how to interpret the results of the respective simulations and their range of validity and reliability.
- *Response–impact research*: Because—as a rule—competing courses of action and alternative strategies are proposed by different societal actors and are based on different scientific conceptions and normative presuppositions, *ex ante* comparisons of these different proposals have to be conducted prior to any decisions. Examples for components of such competing action strategies are conceptual and institutional innovations, improvements in technical efficiency (Mappus, 2005), influencing consumption in the direction of increased sustainability, and introducing new and more sustainable production patterns in industry

(Weaver *et al.*, 2000). Effects of the measures proposed (goal attainment, possible side effects, conditions of implementation) have to be estimated and evaluated in advance.

- *Verification of measures and monitoring their effects*: Measures have to be monitored in the sense that their effects have to be traced and then compared to the initial goals and expectations. This process allows for adapting, modifying and optimizing the measures, but also does not exclude abolishing or reversing them. In order to make learning possible, the causes of deviations of the 'real' effects from the expectations held *ex ante* have to be investigated, in particular. This learning strategy both makes it possible for modified or new measures to be developed and increases knowledge about the system under consideration.

While some of the problems that arise in connection with such research might be solved by progress in the field (e.g. improved modelling of societal activities), other difficulties require new forms of knowledge, such as those discussed in the debate about post-normal science (Funtowicz & Ravetz, 1993) and about mode 2 science. The results of the activities listed above can be distinguished with regard to their different epistemic and pragmatic status, according to the types of knowledge: explanatory knowledge, orientation knowledge and action-guiding knowledge (Grunwald, 2004).

- *Explanatory knowledge*: Sufficient insight into natural and societal systems and knowledge of the interactions between society and the natural environment are necessary prerequisites for successful action on sustainable development. Explanatory knowledge is the knowledge whose creation—according to the classical concept of science—is the specific object of the sciences.
- *Orientation knowledge*: The appraisal of societal circumstances and developments, of global trends and of measures requires *orientational criteria* that permit these factors to be differentiated into 'sustainable' and 'non-' or 'less sustainable' in a comprehensible and transparent fashion. These criteria are *normative* in nature: they are not derived from any observational or experimental study of phenomena, but require a justification which operates with normative premises. According to the respective concept of sustainability, various patterns of argumentation come into question. Different assumptions about the substitutability of natural capital ('strong' vs. 'weak' sustainability) lead to different criteria, as do differences in the normative foundation (for reference to the finite carrying capacity of natural and societal systems versus the justice-theoretical basis).
- *Action-guiding knowledge*: Diagnoses about sustainable development can be provided by the combination of explanatory and orientation knowledge. This is a necessary point of departure for any kind of active steps towards sustainability. Also necessary, however, for any informed decision making regarding sustainable development is foresighted knowledge of measures and of their effects. Action-guiding knowledge comprises answers to the questions as to how therapies can be applied to sustainability weak spots and how implementation problems can be tackled. It includes the development of strategies to convince and motivate actors and to induce societal change. Some of this knowledge consists in transparently revealing the uncertainty and incompleteness of the knowledge, and in indicating courses of action under uncertainty.

Strategic knowledge for sustainable development necessarily consists of combinations of these three types of knowledge: (1) explanations of cause-and-effect chains provide the cognitive basis for every sort of action; (2) orientation criteria are equally indispensable for diagnosis and therapy; and (3) in knowledge for action, they combine in a specific way that provides answers to the question about how to act given specific knowledge and specific orientation criteria.

Focus: Shaping Technology for Sustainable Development

The overall challenge of guiding change toward sustainable development and the role of the uncertainties involved in this will be analysed in this section by taking a closer look at technology development. Despite some specifics, most of the conclusions and results of this examination can also be taken to be valid in other areas, such as politics or business strategies.

The development, production, use and disposal of technical products and systems have an impact on the ecological, economic and social dimensions of sustainable development and vice versa (Mappus, 2005; Weaver *et al.*, 2000). Technology influences key factors concerning the sustainability of human society, such as long-term *environmental consequences* (renewable and non-renewable resources, emissions, etc.), *economic* aspects (e.g. empowering the economies of developing countries) and *social aspects* (such as equity or social integration).

To be able to shape technologies in favour of sustainable development, the assessments of the sustainability of the technologies under consideration have to be performed in the course of their development and *in advance* of the relevant decision making. *Prospective knowledge* on the sustainability effects of the production, use and disposal of these technologies has to be made available. In order to provide such knowledge, a good understanding of the dynamics and the future development of the affected subsystems of society and the respective role of technology is necessary.

The remainder of this section will consider the following sources of uncertainties of knowledge in this field:[3] (1) the inseparability of technical and societal factors influencing sustainability; (2) the need for life-cycle analyses leading to the well-known problems related to predicting sustainability aspects emerging from the use of technology and from user behaviour; (3) the problem of ensuring that all of the relevant sustainability issues have been taken into account; (4) from problems induced by the inevitable transfer of knowledge; and (5) from the challenge of integrating the large manifold of partly incommensurable aspects of sustainability into a coherent picture.

The Inseparability Issue

In order to assess new technologies in terms of sustainability, it is essential that the sustainability effects be determined only partly by the technology alone. Technology, as such, is neither sustainable nor unsustainable. The contributions of technology to sustainability are determined only partially by technical parameters; the ways in which technologies are used and embedded into society exert a strong influence (Fleischer & Grunwald, 2002; Sorensen & Williams, 2002). This implies that, in order to assess the effects of a specific new technology on sustainability, the changes in societal behaviour that correspond to the specific manner in which the respective technology is embedded in society have to be taken into account. Sustainability assessments of technology have to take into consideration include societal processes, structures, values, customs, etc. that might be affected by the manner in which the technology is embedded in society. This means that the effects of technology on sustainable development also include a—possibly extended—set of societal aspects. The effects of technology on sustainability cannot be assessed by looking only at the technology itself, at its features, at the way it will be produced or at the resources needed and emissions produced during the time of its usage. The inseparability of technology from society clearly leads to uncertainties in the knowledge created because of the systemic nature of societal interaction with new technologies, for example, as a result of feedback loops and the emerging effects of self-organization (Voss *et al.*, 2006). Such systemic effects (e.g. rebound effects) could undermine the conditions for the desired positive effects on sustainability and could lead to the emergence of unexpected negative consequences.

The Life Cycle Issue and the Prediction Problem

Shaping technologies for sustainable development requires *anticipatory* assessments of sustainability to make it possible to distinguish between more or less 'sustainable' technologies (Fleischer & Grunwald, 2002; Sorensen & Williams, 2002). Such anticipatory assessments are concerned with the production phase, the usage phase and the disposal of technical products and systems. They have to cover the entire *life cycle* of a technological product or system. The life cycle starts with exploring and mining natural resources and raw materials, leads via transport and various treatment processes to the manufacturing areas where components of a system are produced, extends to the—intended and possibly unintended—use of the technology in society, the impacts and consequences of the use for both the natural environment and society, and finally must take into account the disposal of these products. Sustainability assessments of technology must consider any effects on sustainability over time, especially the net effect on sustainability during the complete life cycle.

Because the sustainability assessments of new technologies shall contribute to shaping those technologies and, therefore, must be performed in early stages of the development, from the perspective of technology designers and decision makers seeking orientation toward sustainable development, the life cycle of new technologies lies in the future. This means that 'prospective life cycle analyses' are needed. They have to deal with the problem of conceptualizing and assessing future developments: consumer and production patterns, future developments of lifestyles and markets, political and economic boundary conditions for the later usage of new technologies are just some examples of aspects of the future which should be known *in advance* in order to provide decision-guiding life-cycle analyses as part of sustainability assessments.

Many arguments against the predictability of societal issues and of the development of complex socio-ecological systems have, however, been identified by many authors. As a consequence, the technology assessment (TA) community today regards the future as an open space which could be addressed and structured only in the form of 'possible future'-like scenarios rather than be predicted in the older approach of prognoses. If the general idea of life cycle analysis is to be taken as an element for assessing the sustainability of new technologies—and there are good reasons in favour of this approach—then this approach itself would have to be modified considerably. The methodological challenge will be to enable *prospective* analyses to consider life cycles that include uncertain facets.

Taking seriously this point and the inevitable uncertainties of *prospective state-ments*—independent of whether they concern societal states, processes or the effects of guidance measures on sustainable development—has a dramatic impact on the shaping scheme: while predictions are used in classical guidance regimes (cf. Camhis, 1979) to describe the (assumingly fixed) boundaries within which the goals have to be approached in a most efficient way, the outcome is that no or only few such boundary conditions exist with regard to working towards sustainability.

A fortiori it might even be the case that the implementation of measures in favour of sustainable development could challenge the conditions that had pre-viously been regarded as relevant for the success of the measures. This is a further inseparability issue at the heart of the challenge to working towards sus-tainable development: the actors and decision makers cannot separate some unaf-fected boundary conditions—which might be reliably predicted—from the fields they want to influence and to shape by applying certain sustainability measures. There is no external observer's standpoint from which such a separation could be carried out. All of the actors are—of course, in different roles—*participants* in common transformation processes. This will give rise to further reflection.

The Completeness Issue and the Incompleteness Problem

The normative dimension of sustainability shows many different and hetero-geneous aspects, ranging from dealing with natural resources to organizing society around postulates of equity and justice (Grunwald & Kopfmüller, 2006). Two different challenges for steering arise from this situation: first, many aspects cannot be defined unambiguously by scientific analysis but are value-laden, contested and the subject of conflict and will, therefore, not be analysed in this paper. The second point, namely, the complexity of sustainability and the resulting uncertainties, does, however, fall within the scope of this paper.

Many fields have to be observed, and a large number of indicators have been proposed in order to fully picture the empirical societal situation of sustainability (Kopfmüller *et al.*, 2001; Opschoor *et al.*, 1991). However, there can be no guarantee that this specific 'full picture' will really be derived from the assessment. Relevant sustainability aspects might be simply overseen or excluded from further con-sideration after being erroneously deemed of low relevance in that field. It is impossible for philosophical, economic and pragmatic reasons to gain a complete picture in investigating sustainability aspects of a new technology. Decisions have to be made on the relevance or irrelevance of specific dimensions, and the limit-ations of the systems considered. Such decisions, however, are risky in them-selves. It will often be only clear after decision making, production and use of that technology whether these were based on correct judgements—but then it might be too late to change anything (Collingridge, 1980). The risk of incomplete knowledge in decisions on the assumed sustainability of new technologies is una-voidable, and will generally lead to uncertainties in the process of decision making and governance.

The Transfer of Knowledge and Validity Problems

In assessing the sustainability of new technologies, some knowledge must be transferred to different areas. In particular, knowledge gained from simulation based on specific models of technology development, diffusion and use, and including, for instance, models of customer behaviour or competition in that field, must be transferred to the 'real' world. A main assumption is that the basic models were established 'adequately' in the research and assessment of the sustainability impacts of that technology. There is, however, no advance proof available whether this assumption is correct. A certain risk of not meeting the real world's developments remains and leads to uncertainty in applying the results of such assessment to decision making and steering.

The Integration Issue and the Incommensurability Problem

The same initial conditions—many different and heterogeneous aspects of sustainability, ranging from managing natural resources to organizing society in terms of postulates of equity and justice and thereby observing many completely different indicators (cf. Grunwald & Kopfmüller, 2006)—also lead to a different issue: the heterogeneity of the various ethical dimensions and indicators of a technology's sustainability means no common measure of sustainability can be applied to them as a whole. In terms of method, it not satisfactory to measure emissions of greenhouse gases, numbers of people affected by long-term unemployment, information on co-operation with developing countries, or the existence of civil society organizations and engagement according to the same unique scale. There is no chance to integrate the various dimensions and indicators of sustainable development into one single measure like a 'sustainability index' of a technological system or product without incurring severe methodological problems. As the sustainability imperative is integrative by nature, any attempt to integrate data or assess results from heterogeneous aspects of 'overall' sustainability demonstrates normative dimensions which strongly limit the use of common decision-analysis tools. Rather, discursive tools must be used, and political decisions made (Skorupinski & Ott, 2000; Shrader-Frechette, 1991).

This 'incommensurability issue' does not affect the usual set of knowledge in the field, but leads to specific uncertainties of *integrated* knowledge. It might be particularly the case that both positive and negative contributions to sustainable development might arise from the same technology but with regard to different sustainability indicators or aspects. If, in this case, there is no means of integrating the diverging results into a quantitative balance and then deciding which contribution is dominant, the assessment exercise will be left up to careful and qualitative consideration as well as to complex processes of weighing arguments and defining priorities—which, again, makes the use of discursive tools necessary. This situation restricts the options for using quantitative methods to a considerable extent and, if some of them are used, such as cost–benefit analysis or MCDA, their results have to be integrated with other results from more qualitative argumentation and investigation—which is, in terms of method, a delicate endeavour.

Conclusions

Technology can be assessed with regard to sustainability only under conditions of uncertainty and incomplete knowledge. As a result, the notion of 'sustainability potentials' of innovative technologies is currently being widely used. This notion seems, at first glance, to avoid the above-mentioned problems of predicting the sustainability effects of new technology developments. However, there are at least two severe problems.

(i) It would be hard to imagine a technology for which no sustainability potentials at all could be identified. But potentials are *only potentials*—which means that they will neither automatically become reality at some time nor that is there any clear statement on the probability or conditions of their becoming reality. Talking about potentials is akin to making a promise without knowing whether it will be realized or is even feasible. Further knowledge (a certain type of orientation knowledge, see above) would be needed in order to enable decision makers to assess the validity of the promises.

(ii) It should not be forgotten that—besides the positive potentials of new technologies for sustainable development—there might also be *negative* effects. These are also hypothetical, but as soon as positive potentials are used as a replacement for a sustainability assessment, the possible negative effects should also be taken into account to provide a comprehensive and balanced picture—even if this picture too remains uncertain.

Independent of the approach and ideas chosen, the uncertainty issue makes it impossible to foresee the sustainability effects of a technology in the development phase completely and with certainty. Uncertainties always remain. *This situation requires an adaptive approach to shaping technologies in favour of sustainable development*, including permanent monitoring of new knowledge as it becomes available and of integrating new knowledge and new orientations even in the later phases of development—which is, of course, often difficult.

Uncertainties of Knowledge—Further Analysis

The arguments presented above for shaping technology towards sustainable development more or less indicate more general characteristics of approaching sustainability by guidance measures. The *inseparability issue* is relevant for each governance activity towards sustainable development. In prospectively assessing the consequences and impacts of implementing certain measures, not only the field that will be influenced must be considered but also others which might be affected. An assessment of sustainability measures must not be restricted to separate parts of society. If such a restriction has to be made for pragmatic reasons (which is usually the case), then the export/import effects at the border of the system considered must be investigated carefully. The *incompleteness issue* may be transferred to more general shaping activities without any change in its meaning. This also holds for the *incommensurability issue, the transfer issue* and for the *prediction issue*. In order to get a better idea of how to deal with these issues relating to uncertainty of knowledge, it seems to be appropriate to explore some aspects of uncertainty and its relation to sustainability governance more thoroughly.

Origins of Uncertainty

There are different types of origins, sources and causes of uncertainties that are relevant to sustainability governance (in general, cf. Gottschalk-Mazouz &

Mazouz, 2003). Without giving a comprehensive picture, only the following types shall be mentioned here (cf. Pellizzoni, 2003; Walker & Harremoes, 2003):

- origins at the *data level*: uncertainties arising from imperfect or incomplete data sets;
- origins at the *dynamics level*: (i) incompleteness of knowledge about systems dynamics in various fields, especially in coupled systems of nature/society (Rotmans, 1999); (ii) the inseparability of sectoral developments from the development as a whole with the consequence that import/export relations must be considered—resulting in a much higher complexity and the need for knowledge integration which carries its own uncertainties; (iii) many more recombination possibilities of particular effects, of spreading uncertainties through the whole system, of unforeseeable combination effects of different uncertainties, and of the emergence of unexpected systemic effects (such as rebound effects);
- origins of *extrapolation to the future*: sustainability assessments of guidance measures depend on assumptions on future developments in society and at the interface between society and the environment with their inevitable uncertainties (cf. third section).

These various origins of uncertainty are different in nature and have to be dealt with in different ways. Some are ontological in nature and, therefore, indispensable while others might, at least partly, be overcome by improved methods and models, or by more and enhanced empirical research.

Towards a Theory of Future Knowledge

Voss *et al.* (2006) emphasized the fact that uncertainties must be differentiated in terms of *degree*. Going one step further, it seems that what is missing is a 'theory' of (uncertain) future knowledge. Philosophy of science has invested a lot of work in epistemology with the claim of certainty or even truth. The issue of uncertainty, however, has been dealt with in only a few cases (cf. Funtowicz & Ravetz, 1993; Gottschalk-Mazouz & Mazouz, 2003).

Establishing a theory of future knowledge is more complicated than expected at first glance. It would consist of analysing the various types of knowledge necessary to assess sustainability in terms of their claim to validity (*Geltungsanspruch*) and the options for realizing that claim by providing sound arguments and scientific evidence. There are a number of questions arising in this field.

- What can be said about the epistemological status of knowledge in the field involved compared with the usual standards of scientific knowledge?
- How and under what conditions can the *extrapolation* of such knowledge to the future be argued for and justified? Are there counterarguments and how can both types of argument be weighed against each other?
- How do the projected developments depend on future intentional societal decision making and on assumptions about evolutionary developments of human behaviour?
- What role do mere *ad hoc* assumptions play in future projections and how could one 'rationalize' them?

The aim should be—but this would go far beyond the scope of this paper—to establish a typology of knowledge types with different degrees (and origins) of uncertainty. Besides theoretical interest in such an epistemology it would be highly relevant to the problems of working towards sustainable development. Especially because 'contested futures' (Brown *et al.*, 2000) play an important role in the debates on sustainability measures, it seems to be urgent to establish tools for assessing such diverging pictures of the future. Tools to this end will be, on the one hand, analytical ones because the epistemological and normative content and premises of the futures should be uncovered in a transparent way. On the other, discursive tools are needed because of the ethical incommensurability of the criteria which must be taken into account in the assessment. Some first ideas on a procedural 'Foresight Knowledge Assessment' have been identified by Pereira *et al.* (2007) and, in the particular field of assessing future visions, by Grunwald (2007).

On the Relation Between Guidance Actors and Systems

A far-reaching distinction can be identified in many approaches to scientific modelling and simulation of systems relevant to sustainability, namely, the distinction between the system to be influenced by guidance measures and the actors who are responsible for their implementation. The basic assumption is that the observer of a system may be decoupled from the system observed.[4] The question, however, is whether the basic assumption that an external observer's or governor's standpoint is available at all still holds for such complex issues like working towards sustainable development. Under which conditions does the classical distinction between an observer and the system observed still work and how could the criteria for assessing the respective type of situation be established?

There are several indicators that this will, in general, not be the case. In recent work on innovation processes and governance issues (Voss *et al.*, 2006), the main governing agents are regarded as part of the system rather than external forces building on earlier work (e.g. Willke's (1983) emphasis on 'contextual steering'). In the context of this paper, there is a specific argument for this 'internal' view of guidance. In using predictions and future projections for orientating sustainability measures, one can sometimes observe issues like self-fulfilling or self-defeating prophecies. These phenomena result from the situation that the predictive statements made are not simply the results of external observations but at the same time change the boundary conditions for societal groups and actors. In this way, the predictions will influence and modify the conditions that have been assumed to be valid while the prediction was being established. The predictor in these situations is not a distant observer but simultaneously a participant in ongoing communication who might—by his or her observations and the resulting predictions—change the acting conditions for other people. Here, one sees that the distinction between a distant observer who does not intervene and the system to be observed and guided does not work in complex fields like sustainability policy.

In extreme cases, it might come to paradox situations where attempts to reduce uncertainty could lead to an increase in uncertainty. For example, if research activities were initiated to provide knowledge on life-cycle data of specific materials, the aim would be to create knowledge and to reduce uncertainty in order to perform 'valid' sustainability assessments. However, the research might end in ambivalent situations. Unexpected effects might be identified which gave rise to the emergence of rebound effects. Because the real impact of such rebound effects would depend on future consumer behaviour, which is only predictable to a small extent, the sustainability assessment would end with less uncertainty in some fields but more in others. There is no guarantee of reducing uncertainty by answering research questions, because research and assessment might also create new questions, new challenges and new uncertainties.

Uncertainty As a Chance?

Frequently, the issue of uncertainty is discussed in a 'lamenting' tone: the uncertainty of knowledge endangers the design and implementation of promising sustainability measures. It hinders activities working towards sustainable development and might even be misused by politicians as an argument for a 'wait and see' strategy: if the knowledge is uncertain, some might say, then we should wait until science has done its homework and can provide certain and valid knowledge so that guidance towards sustainable development can be done in the same way as mechanical or socio-technological engineering. Scientists also often approach uncertainty as something which should generally be overcome. However, there is no way of gaining certain knowledge about future developments. In all attempts to decrease uncertainty, one should accept that there are limitations and that uncertainties will remain in spite of the scientific effort to overcome them.

Complementing this epistemological line of argumentation, there is an additional—ontological—argument. While the epistemological arguments focus on the impossibility of achieving certain knowledge about the future because of the limitations of our epistemological tools, the argument of a principal indeterminacy and openness of the future states that the future is *per se* not determined and, therefore, not predictable, regardless of the capabilities of our epistemological toolbox. The reason can be seen, for instance, in basic statements of philosophical theory of action, in assumptions about the openness of discursive deliberation in preparing decisions and actions, or on other anthropological or sociological grounds. Independent of the underlying argumentation, it holds that openness of the future and the possibility of shaping future developments are two sides of the same coin.

However, one must observe that there is a very general difference between the preliminary nature of (scientific) knowledge, which always allows for improvement, modification, even falsification and complete revision, on the one hand, and the limited revisability or reversibility of actions that have already been taken. For example, in technology development and implementation, there are economic and practical reasons restricting the possibilities of adjustments and modifications at later stages of the development. As soon as a highway has been built, it will persist even if a sustainability assessment showed its disastrous impacts. Consequently, there are limitations to the openness of the future.

Nevertheless, even considering this point, there is no reason for merely 'lamenting' about our incapability of prediction. Instead, one could try to regard uncertainty as a *chance* as far as possible. Uncertainty—besides its negative connotations of having no certain knowledge for planning and decision-making processes—denotes the *openness* of the future, which is not an openness in an arbitrary sense but in the sense that shaping activities can and will have an impact on the further course of development. It allows for learning processes, for working according to self-defined goals, for deliberating about how to shape the future and for proactively thinking about future society. Therefore, uncertainty comprises not only risks for directing activities but simultaneously opens up spaces for reflection, learning and open deliberation.

Uncertain Knowledge and Guidance Action

The discussion in the third section and the classifications and analyses in the fourth section showed concurrently that working towards sustainable development is structurally far removed from the classical planning situation (Camhis, 1979). Instead, the earlier work of Lindblom (1973) can be built upon and his work interpreted in the sense of a "directed incrementalism" (Grunwald, 2000). In this chapter, some consequences for guidance action will be addressed.

Evidence Thresholds to Motivate Precautionary Action

A frequently discussed question is whether and to what extent the knowledge currently available is sufficiently certain and complete to legitimize sustainability-relevant action, for instance, in view of the societal costs connected with it. The demand for complete knowledge, or the demand for guaranteed certain knowledge can paralyze action or even be (mis-)used to delay or prevent sustainability-relevant measures: as long as the way the systems work is not known completely and with certainty—thus runs an opinion occasionally heard—it is impossible to decide on the necessity and aptness of measures. This position would take the uncertainty and incomplete nature of knowledge as an argument for doing nothing—which, in many cases, could lead to the risk of waiting too long: it could be too late to prevent catastrophes or irreversible unsustainable developments (see Harremoes *et al.*, 2002 for case studies).

Of course, precautionary strategies must not be overstressed. A certain 'evidence threshold' must be exceeded, if drastic measures are to be justified. Not every mere suspicion (for instance, in the sense of a "heuristics of fear" or of a "primacy of the negative prognosis" (Jonas, 1979)) leads to massive political and societal regulatory measures; this would cause a complete blockade. In order for the precautionary principle to become effective—and sustainability is inseparably bound up with precaution—a potential hazard has to be scrutinized comprehensively. The many debates on how to identify a balanced approach in applying the precautionary principle and analysing its impacts have their epistemological and ethical origin in the difficulties involved in dealing with the uncertainty, incompleteness, or even the absence of knowledge (von Schomberg, 2005).

The Experimental Nature of Sustainability Policies

The uncertainty and incompleteness of our knowledge (together with the ambivalence of the goals) make it impossible to pursue sustainability by means of "rational comprehensive planning" (Camhis, 1979). A policy of sustainability thus has to be implemented under conditions of uncertain knowledge and of provisional assessments. *Ex ante* one cannot know for certain whether and to what extent a political measure, a technological innovation, or a new institutional arrangement will 'really' contribute to sustainability when actually applied. Every sustainability policy has to confront this situation and become—in a certain sense—'experimental' (Braybrooke & Lindblom, 1963).

In this experimental situation, it is crucial for this uncertainty and incompleteness in our knowledge not to paralyze or hinder action, but to seize the widest range of opportunities for *learning* in implementing practical measures and to avoid irreversible effects as far as possible. The management of the implementation of such measures remains experimental—even if excellent *ex ante* assessments would be available. As in an experiment, it is necessary to monitor the real effects carefully and to be prepared to modify or replace the implemented measures. In this sense, the road to the future is incremental (Lindblom, 1973) and must allow for learning during the series of incremental steps—which can always be evaluated against the normative sustainability framework that safeguards, to a certain extent, the 'direction' of the process (Grunwald, 2000). The Leitbild of sustainability does not provide a fixed and operable set of criteria but rather a discursive framework full of ambivalences and conflicts. None the less, it includes a normative kernel, consisting of the ideas of responsibility for the future and of equity (Grunwald & Kopfmüller, 2006), which serve as a compass in at least some sense.

The practical application of political measures *over a certain span of time* therefore offers an essential opportunity for gaining the knowledge that was unavailable in the preparatory phases, including anticipatory sustainability assessments, in order to be able to modify and optimize the measures accordingly. The task of empirical research is, therefore, to observe the effects of the implemented measures (monitoring) to allow for improvements and readjustments in the measures.

Directed Instead of Disjointed Incrementalism

In opposition to the rational comprehensive planning approach (cf. Camhis, 1979 for a detailed reconstruction), Lindblom and others developed a more incrementalistic, 'muddling through' approach (Braybrooke & Lindblom, 1963; Lindblom, 1973), which has been termed "disjointed incrementalism" (Camhis, 1979). Some of the arguments in favour of this approach are analogous to those given above, namely, that working towards sustainable development is not possible according to a classical planning approach. However, the direct transfer of the ideas of 'disjointed incrementalism' to our challenge of approaching sustainable development does not work either.

Sustainable development implies that societal processes of development—including technology development—are re-orientated so as to ensure that the needs of future generations and those of the Southern Hemisphere are taken into account as well as those of current generations in the Northern Hemisphere. Thus, sustainable development necessarily includes dealing with long-term and normative considerations. It involves taking into account (normative) aspects of the distant future, of the impact of our present concepts of technology and society on this future, and the impact of such reflections on our present-day concepts and ideas (*backcasting*). Proceeding in a purely incremental way would not allow us to meet these requirements, because disjointed incrementalism does not foresee normative and longer-term guidance elements.[5]

Because of this, it has been proposed that disjointed incrementalism be transformed into a "directed incrementalism" (Grunwald, 2000). This approach would allow the direction of action and decision towards sustainable development to be maintained without fixing goals in detail or falling back on a traditional planning approach. Permanent reflection on the goals and the means to attain them according to the results of continuous monitoring of ongoing developments could lead to incremental changes in the direction of the development, in the goals and in the measures to meet these goals. This model allows us to deal constructively with the uncertainties involved in continuous learning processes, reviews of sustainability assessments and adjustments of measures.

Reflexive Sustainability Policies and the Need for Meta-Knowledge

The situation described in the previous section implies a considerable increase in the demands on reflexivity in the sustainability sciences as well as in policy making for sustainability. Two types of reflexivity may be distinguished, both of which are required to allow guidance towards sustainable development while facing the situation of wide-ranging and unremovable uncertainties.

(i) *Iterative science-internal reflection* on the premises of its own research and the sustainability assessments provided is required. Two different levels of knowledge both have to be taken into consideration. If strategic knowledge for sustainable development (cf. second section) is classified as knowledge of the first order, *then knowledge of the second order consists of knowledge of the conditions for the validity of knowledge of the first order.* Normative premises, cognitive presuppositions, the limits of the system observed, judgements of relevance and awareness of the epistemological limitations of knowledge of the first order belong in this category, as does knowledge of the inherent uncertainties. In addition to the strategic knowledge for sustainable development produced by sustainability sciences, a *meta*-knowledge must therefore also be provided, which includes the results of normative and epistemological reflection. Society and policy makers must not only be provided with action-guiding knowledge, but also with cognizance how to interpret this knowledge, and where its limits lie. This type of reflexivity is the prerequisite for science's capacity to further develop its own sustainability assessments on the basis of new data and knowledge as well as to changing normative backgrounds.

(ii) *Institutionalized societal learning processes involving the relevant societal groups should increase societal reflexivity.* According to the necessity identified for monitoring the 'real' impacts of guidance measures in order to enable society and policy makers to modify, to adapt, to reverse or to further develop these measures, learning processes have to relate knowledge production (in the sense of the above-mentioned reflexive approach) to action and decision making. Because of the uncertainties involved, working towards sustainable development does not consist of a linear application of measures. Instead, it comprises multi-step learning cycles which, in a first step, lead from knowledge production and provisional sustainability assessments based on that (in part uncertain) knowledge to guidance actions (cf. van Asselt & Rotmans, 2001). In a second step, the effects of these actions have to be monitored in the 'real' world. In the third step, the results of this monitoring would be analysed with regard to sustainability goals, and refinements of the measures would be applied. In this reflexive way, combining scientific reflexivity with societal reflexivity, and including both first-order and second-order reflexivity, sustainable development could be approached in the form of "directed incrementalism" (Grunwald, 2000).

Acknowledgement

The author would like to express thanks to Jan-Peter Voß, Jens Newig, Jochen Monstadt and two anonymous referees for many valuable remarks and recommendations that have contributed considerably to improving this paper.

Notes

1. All of the analyses and arguments are presented here at a conceptual level. References to specific cases are made for purposes of illustration only, where appropriate.
2. The concept of knowledge used throughout this paper is a rather general one that is far from restricting knowledge in a positivistic sense. To categorize a statement as knowledge, however, is not arbitrary but bound to standards which might differ in various areas. But, in any case, using the term 'knowledge' involves the regulative idea that it should be possible to prove its reliability or validity (e.g. by scientific means or by discursive procedures) (Luhmann, 1990).
3. There are also other sources of uncertainty which relate directly to the social and economic processes of technology development, for example the interdependence between technologies and their path-dependence.
4. This has been the classical epistemological situation in science. Transferred to the steering challenge, this relates to the distinction between governing a system from inside or from outside, cf. Smith & Stirling (2007).
5. The kernel of incrementalism consists of the trial-and-error approach, as suggested in some incrementalistic planning theories (Braybrooke & Lindblom, 1963; cf. the discussions in Camhis, 1979).

References

Braybrooke, D. & Lindblom, C. E. (1963) *A Strategy of Decision* (New York: The Free Press).

Brown, N., Rappert, B. & Webster, A. (2000) (Eds) *Contested Futures. A sociology of prospective techno-science* (Burlington: Ashgate).

Brown-Weiss, E. (1989) *In Fairness to Future Generations. International Law, Common Patrimony and Inter-generational Equity* (New York: United Nations University).

Camhis, M. (1979) *Planning Theory and Philosophy* (London: Pinter).

Collingridge, D. (1980) *The Social Control of Technology* (New York: Tavistock).

Fleischer, T. & Grunwald, A. (2002) Technikgestaltung für mehr Nachhaltigkeit - Anforderungen an die Technikfolgenabschätzung, in: A. Grunwald (Ed.) *Technikgestaltung für eine nachhaltige Entwicklung*, pp. 99–148 (Berlin: Edition Sigma).

Funtowicz, S. & Ravetz, J. (1993) Science for the post-normal age, *Futures*, 25(7), pp. 739–755.

Gottschalk-Mazouz, N. & Mazouz, N. (Eds) (2003) *Nachhaltigkeit und globaler Wandel. Integrative Forschung zwischen Normativität und Unsicherheit* (Frankfurt&New York: Campus).

Grunwald, A. (2000) Technology policy between long-term planning requirements and short-ranged acceptance problems, in: J. Grin & A. Grunwald (Eds) *Vision Assessment: Shaping Technology in 21st Century Society*, pp. 99–148 (Heidelberg: Springer).

Grunwald, A. (2004) Strategic knowledge for sustainable development: the need for reflexivity and learning at the interface between science and society, *International Journal of Foresight and Innovation Policy*, 1(1/2), pp. 150–167.

Grunwald, A. (2007) Converging technologies: Visions, increased contingencies of the conditio humana, and search for orientation, *Futures*, 39, pp. 380–392.

Grunwald, A. & Kopfmüller, J. (2006) *Nachhaltigkeit* (Frankfurt & New York: Campus).

Harremoes, P., Gee, D., MacGarvin, M., Stirling, A., Keys, J., Wynne, B. & Guedes Vaz, S. (Eds) (2002) *The Precautionary Principle in the 20th century. Late Lessons from early warnings* (London: Earthscan).

Jasanoff, S. (2004) Ordering Knowledge, Ordering Society, in: S. Jasanoff (Ed.) *States of Knowledge: The Co-Production of Science and the Social Order*, pp. 43–68 (London: Routledge).

Jonas, H. (1979) *Das Prinzip Verantwortung* [*The Imperative of Responsibility*] (Frankfurt: Suhrkamp) (translation published in 1984).

Kates, R. W., Clark, W. C., Corell, R., Hall, J. M., Jaeger, C., Lowe, I., McCarthy, J. J., Schellnhuber, H.-J., Bolin, B., Dickson, N. M., Faucheux, S., Gallopin, G. C., Gruebler, A., Huntley, B., Jäger, J., Jodha, N. S., Kasperson, R. E., Mabogunje, A., Matson, P., Mooney, H., Moore, B., O'Riordan, T. & Svedin, U. (2000) Sustainability Science, *Science*, 292, pp. 641–642.

Kopfmüller, J., Brandl, V., Jörissen, J., Paetau, M., Banse, G., Coenen, R. & Grunwald, A. (2001) *Nachhaltige Entwicklung integrativ betrachtet. Konstitutive Elemente, Regeln, Indikatoren* (Berlin: Edition Sigma).

Lindblom, C. E. (1973) The Science of 'Muddling Through', in: A. Faludi (Ed.) *A Reader in Planning Theory*, pp. 151–170 (Oxford: Pergamon Press).

Luhmann, N. (1990) *Die Wissenschaft der Gesellschaft* (Frankfurt: Suhrkamp).

Mappus, S. (Ed.) (2005) *Erde 2.0 – Technologische Innovationen als Chance für eine Nachhaltige Entwicklung?* (Berlin: Springer).

Nowotny, H., Scott, P. & Gibbons, M. (2001): *Re-Thinking Science. Knowledge and the Public in an Age of Uncertainty* (Oxford: Polity).

Opschoor, H. & Reijnders, L. (1991) Towards Sustainable Development Indicators, in: O. Kuik & H. Verbruggen (Eds) *In Search of Indicators of Sustainable Development*, pp. 7–27 (Dordrecht: Kluwer).

Pellizzoni, L. (2003) Knowledge, Uncertainty and the Transformation of the Public Sphere, *European Journal of Social Theory*, 6(3), pp. 327–355.

Pereira, A. G., von Schomberg, R. & Funtowicz, S. (2007) Foresight Knowledge Assessment, *International Journal of Foresight and Innovation Policy* 4(1), pp. 65–79.

Rotmans, J. (1999) Global Change and Sustainable Development: Towards an Integrated Conceptual Model, in: H.-J. Schellnhuber & V. Wenzel (Eds) *Earth Systems Analysis. Integrating Science for Sustainability*, pp. 421–450 (Berlin: Springer).

Schellnhuber, H.-J. (1999) Earth Systems Analysis – The Scope of the Challenge, in: H.-J. Schellnhuber & V. Wenzel (Eds) *Earth Systems Analysis. Integrating Science for Sustainability*, pp. 3–195 (Berlin: Springer).

Sen, A. (1987) *On Ethics and Economics* (Oxford: Oxford University Press).

Shrader-Frechette, K. (1991) *Risk and Rationality. Philosophical Foundations for Populist Reforms* (Berkeley: University of California Press).

Skorupinski, B. & Ott, K. (2000) *Ethik und Technikfolgenabschätzung* (Zürich: ETH).

Sorensen, A. & Williams, S. (2002) *Shaping technology, guiding policy* (London: Elgar).

van Asselt, M. & Rotmans, J. (2001): Uncertainty Management in Integrated Assessment Modeling: Towards a pluralistic approach, *Environmental Monitoring and Assessment*, 69, pp. 101–130.

von Schomberg, R. (2002) *The objective of Sustainable Development: Are we coming closer?* EU Foresight Working Papers Series 1 (Brussels: Commission of the European Union)

von Schomberg, R. (2005) The Precautionary Principle and Its Normative Challenges, in: E. Fisher, J. Jones & R. von Schomberg (Eds) *The Precautionary Principle and Public Policy Decision Making*, pp. 141–165 (Cheltenham, UK/Northampton, MA: Edward Elgar).

Voss, J.-P., Bauknecht, D. & Kemp, R. (Eds) (2006) *Reflexive Governance for Sustainable Development* (Cheltenham: Edward Elgar).

Walker, W. E. & Harremoes, P. (2003) Defining Uncertainty. A Conceptual Basis for Uncertainty Management in Model-Based Decision Support, *Integrated Assessment*, 4(1), pp. 5–17.

WCED–World Commission on Environment and Development (1987) *Our common future* (Oxford: Oxford University Press).

Weaver, P., Jansen, L., van Grootveld, G., van Spiegel, E. & Vergragt, P. (2000) *Sustainable Technology Development* (Sheffield: Greenleaf Publisher).

Willke, H. (1983) *Entzauberung des Staates. Überlegungen zu einer gesellschaftlichen Steuerungstheorie.* (Königstein/Ts.: Athenäum).

Risk Management at the Science–Policy Interface: Two Contrasting Cases in the Field of Flood Protection in Germany

HELLMUTH LANGE & HEIKO GARRELTS

Introduction

Due to its potential for risk and disaster, climate change can increase the vulnerability of society. Thus, societies have to pursue strategies encompassing mitigation as well as adaptation and response. Vulnerability to climate variation is determined in part by the politically adopted strategies in force so far. Flood protection is one of the fields in which to cope with the problem means to seize or to miss the respective chance. According to Paavola & Adger (2004, p. 175), adaptive responses include changes in institutional arrangements or public policies, public and private spending, as well as investments in infrastructure and other durable goods. However, experience also shows that there are constraints on achieving full adaptation. Maladjustments may occur due to decisions that are based on short-term considerations, imperfect foresight, insufficient information and over-reliance on insurance mechanisms (IPCC, 2001, p. 8). Another crucial aspect of adaptation is the generation, dissemination and consideration of climate-related knowledge.

With regard to the latter point, the actors inevitably come up against a dilemma: scientific research on climate change is providing an ever-growing amount of knowledge. However, being based on models and notoriously insufficient data, scientific expertise is bound to uncertainty. Thus, the extent, the date of occurrence and the frequency of extreme climatic events can be foreseen only in a rather vague and fuzzy way. Furthermore, scientific scenarios and prognoses are disputed even within the scientific community. Politicians and administrative officers, however, are obviously strongly interested in basing their decisions on the most reliable knowledge available.

This raises the question: what is and what can be the role of science within the context of political and administrative decision making? This paper discusses whether knowledge plays a role and, if so, how this role influences decision-making processes. Which actors refer to what kind of climate-related knowledge? Under what circumstances can new knowledge 'diffuse' into policy, which in turn has to give a public account of the risks of climate change and the inevitably related uncertainties of prognosis. The occurrence of what will be called 'risk discourse'—in opposition to a 'safety discourse'—will be the variable to explain.

In discussing the science–policy interface, reference is made to the debate on blurring boundaries between science and policy making. In earlier models science was seen as delivering instrumental knowledge to policy makers, providing a sound basis for the most adequate political decisions. Based on empirical findings, mainly in the field of science and technology studies, this kind of optimistic (and technocratic) understanding has largely been questioned and, in parts, been replaced by a more sceptical and constructivist understanding (Jasanoff *et al.*, 1995; Weingart, 2001). In fact, science is influencing political decision making increasingly, but science is being politicized increasingly, too. However, blurred boundaries do not imply that the two systems, science and policy, are becoming congruent. Due to the different objectives of science and policy, boundaries remain. Their concrete interaction has to be identified empirically. For this purpose a discourse-analytical framework is employed as a tool to analyse conflicting knowledge claims. These are interrelated with existing narratives on how to cope with the threat of climate change. Two cases are discussed. Both focus on the political–administrative system (administrative officers and political authorities) as the most relevant steering authority.

The first case refers to coastal protection in northern Germany. Within the framework of an empirical research project, the aim was to understand to what degree the more recent scientific debate on climate change has brought about new forms of assessing the necessary dimensions of dykes and further protective structures. In this case, the administrative officers in charge of coastal protection proved to be the key actors, whereas the politicians simply followed and consented to administrative input. Administrative officers were able to play this role due to their specific authority, allowing them to 'reshape' uncertainty to match better with the demands of political and administrative decision making. They developed a "scientific-administrative hybrid" (Potthast, 1999).

The other case study refers to the question of how the political and administrative authorities are coping with the risk of more frequent and more intense flood events in river basins.[1] This project refers to the severe floods in Germany in 2002 and the subsequent policy change towards a more risk-orientated strategy, which is more in line with the emphasis on uncertainty in the ongoing climate-related scientific debate. Thus, this case is an example of steering and 'real' action *despite* incomplete knowledge about the regional impact of climate change (cf. Buchwald, 2008). Policy change clearly resulted from decisions taken by politicians at the federal political level, while the administrations in charge are expected merely to accept and implement these.

Therefore, the focus of either project is on knowledge transfer and/or transformation and policy change, respectively. Whereas the second case led to policy change, the first one did not. Why did the first case lead to developing a scientific–administrative hybrid while the second did not?

The proceeding is as follows. First, a brief overview is given of the complementary tendencies of science being politicized and politics being increasingly influenced by and dependent on science. The two cases are then presented. The findings are interpreted against the background of two different approaches: discourse analysis as emphasizing the constructivist dimension and the policy window approach as emphasizing the importance of political framings and real events. Some conclusions are drawn in the final section.

Conceptual Framework: the Science–Policy Interface

In a functionally differentiated society, policy and science are systems among others. They operate on different logics: the political–administrative system is centred on mediating interests and institutionalizing power in law. Science aims at providing plausible and well-proven explanations. While science aims at testing hypotheses, the political–administrative system is centred on political decision making and on developing subsequent concepts. Political and administrative co-ordination of action is organized in a more or less vertical setting, whereas science is organized in a more horizontal and less integrated way.

However, for quite some time the two systems have been less remote from each other than it may appear in a radical view of system theory. Weingart (2001, pp. 14–15, 25) described how science, since the early nineteenth century, has continuously lost its former social isolation. As a consequence, interdependencies and couplings have been increasing. Thus, the distance between science and the public is shrinking. In 1966, Robert E. Lane dealt for the first time with the 'knowledgeable society': social actors rely increasingly on scientific knowledge, they expend parts of their resources on scientific research, and they use scientific knowledge in an instrumental manner to realize their goals (Lane, 1966, p. 650). Stehr (1994, 36ff.) mentioned further aspects: the diffusion of science into all spheres of life, the replacement of other forms of knowledge, the development of knowledge production as a new sector and the changes in power relations.

The way science gets integrated into political decision making can be described by two different models: a transfer model and a transaction model. Before depicting these models, the aspect of science and uncertainty mentioned in the introduction is discussed.

Science Providing Uncertain Knowledge

Beck (1992) coined the term "reflexive policy". This term aims at precedent political decisions, intended to 'modernize' society constantly in order to ensure a more comfortable life for an ever-growing number of citizens. However, because of an increasing number of more or less unexpected and unwanted side effects, policy making is forced to become reflexive. To what extent is this relevant to knowledge policy as well?

Whereas in the past scientific findings were seen as a 'delivery' to policy makers, today, it is commonly held that scientific findings are temporary and fragmentary. Moreover, systematically produced knowledge not only generates more knowledge but also more ignorance. Against the catchword 'knowledge society' are positioned other catchwords aiming at different forms of not-knowing, in particular 'risk' and 'uncertainty', the latter referring to the lack of prognostic power in open systems (Schiller, 2005, pp. 46ff.).

Uncertainty touches on the authority of natural sciences. This is rather contrary to the expectations of policy makers and to public expectations with respect to science. Weingart (2001, p. 20) concluded: "Linking the knowledge production to decisions in political contexts gives an explosive quality to the problem of ignorance. Scientification of policy loses its primary rationalist sense". How do policy makers deal with this fact?

Transferring Scientific Findings

The idea of directly using scientific knowledge simply by applying it in the ambit of decision making has been influential. It originated in the scientific and technocratic optimism in the field of *policy advising* and science-based *political planning* (Bonß, 2004; Nullmeier, 1993; Weingart, 2001, p. 12). Recent demands on science to provide knowledge that is actively effective in different areas of society are generally made in the context of political challenges and related controversies (Hirsch-Hadorn *et al.*, 2004, p. 285). Discussing the science–policy interface and analysing the scientific input to politics, Norse & Tschirley (2000) adopted the classical policy cycle as a starting point.[2] According to the authors, science provides an inventory of knowledge dealing with the relevant issue and quantifies data for the purpose of problem identification. These serve as a basis for formulating strategies. The next phase leads to a selection of policy options promising to exert a relevant influence on the issue at stake, followed by modelling possible policy implementations, involving physical and economic models. Finally, science also provides the network needed for monitoring and evaluating the respective policy measures (Hirsch-Hadorn *et al.*, 2004, p. 286).

The theoretical background is the idea of knowledge facing policy and the idea of being able to separate facts and values, scientific input, and policy process from each other (Nullmeier, 1993, p. 177). Scientific knowledge, in this perspective, is an external factor with regard to policy analysis, additional to interests, identities or institutions (Scharpf, 2000).

Transforming Scientific Findings

For quite some time now, science has turned to using the media to present its results and to addressing more or less directly its political relevance. By doing so, scientists turn into political actors, thus breaking with the traditional 'scientific ideal' which is the separation of facts and values. Political engagement and even emotions are no longer a taboo. Instructive examples within natural science and environmental policy are the debates and research dealing with biodiversity (Eser, 2001; Takacs, 1996) and the ozone layer (Grundmann, 1999), respectively, and with climate change being of particular interest here (Viehöver, 1997; Weingart *et al.*, 2002). In this perspective, drawing on such issues does not mean to address mere physical phenomena but issues of symbolic meaning, social interest and power as well. This is why, moreover, they can be seen as social constructs, fostered by 'epistemic communities' and shaped by common beliefs and norms (Haas, 1992). One of their basic aims is to influence public opinion and political and administrative actors, in particular.

This kind of (re)defining the science–policy interface is done not only by scientists. As scientific knowledge diffuses into many spheres of society, scientific knowledge in return gives access to a multiplicity of actor groups. In other words, the criteria of quality and relevance concerning science are no longer defined exclusively by scientific actors (Weingart, 2001, p. 15). Hence, objects of science are not simply given, but chosen, interpreted and constructed in accordance with and dependent on general beliefs, interests and social conditions (Jasanoff *et al.*, 1995). This is in line with Nullmeier (1993, p. 177) who emphasized that the notion of externally produced knowledge neglects the capacity of policy actors to produce their own knowledge. They are able to develop their own frames of interpretation and specific 'cognitive representations', according to the *perceived* steering context (Voß *et al.*, 2008, p. 6)

These processes have been called "boundary work" (Gieryn, 1983, 1995). The assertion is that the boundary of what can be considered scientific is neither self-evident nor stable over time. Instead, the boundary between scientific and non-scientific is contested and continuously moved.

Discourses and Policy Windows

At this point, discourses come into play as environmental problems are redefined socially and politically (Arts *et al.*, 2000, p. 60). Transferring or rather transforming scientific findings is one crucial part of giving meaning to environmental problems and designing possible solution strategies. This is done through *discourses*. A policy discourse can be defined as "a specific ensemble of ideas, concepts, and categorisations that are produced, reproduced, and transformed in a particular set of practices and through which meaning is given to physical and social realities" (Hajer, 1995, p. 60). Further, as Dryzek (1997, p. 8) wrote, a discourse "enables those who subscribe to it to interpret bits of information and put them together into coherent stories or accounts". Discourses, in this context, are 'knowledge regimes' which are embedded in scientific practices and techniques and are institutionalized in different policy arenas (Bäckstrand & Lövbrand, 2006, p. 52). This may include either adopting or rejecting uncertainties or risks as constitutive elements of particular policies. As political issues are contested, it is self-evident that discourses also aim at de-legitimizing opposing discourses in order to obtain discourse dominance. Thus, issues get framed in a purposeful manner (Szarka, 2004, p. 318) to provide answers to four W-questions: 'What is the problem?', 'What information is relevant?', 'What has to be done to cope with the problem?', 'Who can do it best?'.

However, changes in discourse dominance depend on the wider political context where policy windows are of particular relevance (Kingdon, 1984). Kingdon emphasized that policy change is far from being a linear process as is implied in the context of the policy cycle approach, assuming a target-orientated, co-ordinated process with well-defined stages. Kingdon was especially concerned with the question why, within public policy, some issues and subjects emerge and are given serious consideration while others remain neglected. Promoting the metaphor of three different 'streams' as an analytical framework, he distinguished a problem stream, a policy stream and a politics stream (public opinion, election, existence or lack of opposition, etc.). He held that each stream is largely independent of the others and that each of them develops according to its own dynamics and rules (Kingdon, 1984, pp. 20ff). The policy stream refers to instruments and conceptions, which float around in a "primeval soup" (Kingdon, 1984, pp. 21, 122ff.). The most significant policy changes occur when all of the three streams (problems, policies and politics) are "coupled into a package" (Kingdon, 1984, p. 21). Kingdon (1984, pp. 173ff.) called this a "policy window": an "opportunity for advocates to push their pet solutions or to push attention to their problems". Policy windows—sometimes predictable, sometimes not—are opened by events in either the problem or the political stream (Kingdon, 1984, p. 213).

The following sections present two case studies.[3] In the first one, scientific findings have been and continue to be *transformed* in order to maintain a safety approach. In the second case, in contrast, uncertainty is the point of departure for a risk approach. Here scientific findings have been *transferred* into policy.

The article will begin by summarizing the two cases. The elements considered in order to explain the different ways of coping with uncertainty and risk are principal actors, institutional settings, discourses, real events and further situational factors.

Two Cases: Concerning Scientific-Administrative Hybrids and Policy Change

The Problem of Flooding

Extreme floods are the most common type of natural disaster in Europe. In flood-prone settlements floods can kill people or make them ill and homeless. They can also damage the environment, infrastructure and property. Most types of floods, such as normal and annual ones, are 'known risks' as they have occurred over thousands of years (Wisner *et al.*, 2004, p. 205). Floods affect some low-lying inland areas because of rainfall, while some coastlines are liable to both rain flooding and sea invasion (especially under storm surge or unusual tidal conditions). Areas at risk are known from earlier events. What is known as well is the role of human activity. Deforestation in mountainous regions accelerates runoff, thereby increasing the likelihood of flooding. Urban development on former flood plains is expected to increase the magnitude of negative impact in the area and to increase the possibility of floods downstream due to the canalization of rivers (EEA, 2004, p. 2).

Because the relevant parameters of flood events are more or less known, there is little uncertainty about the need to take protection measures. In addition, there is little uncertainty about the effect of different policy options (cf. Voß *et al.*, 2007, p. 8). With the given rainfall or storm conditions, both warnings and self and societal protection measures should be possible (Wisner *et al.*, 2004, p. 205).

However, this basic 'expectedness' is reduced by the wide range of intensity and durations of floods that can affect the same area at different times and variation in return periods (the average number of years between the recurrence of floods of a given magnitude). Above all, trends in frequency and intensity of flood events in the future will be related closely to changes in the patterns of precipitation and river discharge and, thereby again, also to other long-term changes in the climate. However, climate change goes along with *prognostic uncertainties* due to general gaps in knowledge, insufficient data availability and difficulties in attributing an observed change to anthropogenic global climate change. Hence, much is known about the climate system and changes in global mean temperature over the past 100 years, but there is only scarce knowledge of climate sensitivity, regional climate change, climate variability and the frequency and intensity of extreme events (EEA, 2004, pp. 82–3).

Coastal Protection on the German North Sea Coast: Between "Safety" and "Risk"[4]

The principal actor is the public administration in charge. Looked at in more detail, this actor proves to consist of several specialized administrative entities.[5] The core assignments of the administration in charge are threefold:

- technical maintenance of the dykes and other protective structures (such as sluices and water barriers);
- assessing and fixing the necessary height and firmness and related constructive features of the protective structures;
- monitoring and integrating external expertise on technical, meteorological and climate issues.

The most relevant point with respect to our topic is how the necessary height and strength of the protective structures are to be assessed.

The procedure in force is strictly empirical. The highest tide gauge ever reported has served and still serves as a reference point from which all other technical data are derived. The only element without clear empirical basis is a safety margin that is added to the respective gauge.[6] Flood occurrence is conceived as swinging within the boundaries of the highest and lowest tides ever reported. Thus, the procedure is seen as a solid basis that allows for ensuring equal safety at all sections of the coastline. As far as changes in the swing itself are concerned, because of climate change, they are—more implicitly than explicitly—assumed to evolve in a continuous and thus linear way. Actually, this way of assessing the necessary dimensions of flood protection has worked fairly well so far: no dyke failures have had to be mourned.

The formula "Equal safety at all parts of the coastline" has become the general guideline for successfully managing coastal protection. On the political level, it was taken on and enacted by law, thus becoming the official mission of the administrative units in charge of coastal protection in Lower Saxony (*Niedersächsisches Deichgesetz*). In the analytical terms of discourse analysis the formula can be called a 'safety discourse'.

However, considering the findings of climate research (see IPCC, 2001, 2007), increasing uncertainty regarding the extent, date of occurrence, frequency and regional specifics of future extreme events have to be envisaged. Against this background, safety proves to be an aim that can no longer be expected to be actually achieved. Instead, risks have to be assessed and determined politically. Risk-related decisions can be seen as decisions following the precautionary principle. In that sense, a 'risk discourse' emerged. The core question is whether and how the administrative officers in charge are ready to shift from the firmly established safety discourse to a risk discourse and whether and which decisions have to be taken in order to reassess and to adjust the system of coastal protection, today or in the near future. Here, it is of particular interest how the actors cope with uncertainty as a core element of climate research.

Unsurprisingly, the interviews provide broad evidence that the administrative officers in charge feel very uneasy with such wide-ranging uncertainty. On the one hand, uncertainty about what will and what can happen does not fit at all with the specific responsibility and professional ethos of the administrative officers. They are in charge of and dedicated to ensuring that the protective structures will be strong enough under all conditions. Therefore, it is a matter of professional self-esteem to be able to ensure that this goal will be achieved in actuality. On the other hand, the scientific debate on climate change, with its emphasis on uncertainty as a constitutive element, cannot be contested in general or even be ignored.

As the safety discourse builds on empirically assessing the required specifications of the protective structures, it is not linked to climate research in any systematic way. In fact, monitoring and assessing the scientific debate, as one of the three core assignments of the administration in charge (assignment 3, see above), initially was only an add-on. Its main purpose was to make sure that new information was assessed continuously in technical respect and, if necessary, adopted for improving the technical quality of the protective structures. Now, faced with the challenge to consider also conceptual readjustments that allow for the emergence of risk instead of continuing to strive for safety, the scientific debate is gaining importance, thus moderately reshaping the responsibilities of the respective divisions with regard to

- assessing the findings of the scientific debate on climate change more systematically and
- filtering and picking up findings of presumed importance for the protection of the German North Sea Coast.

In practice, this tends to boil down to dividing the findings of climate research into two parts: one of relevance and one of no relevance. Basically, there is nothing wrong with that. The important question, however, is, what is the criterion for accepting or rejecting findings of climate research? In fact, there is one general criterion: the degree of certainty associated with data and scenarios.

By looking at the divisions of the administration in charge of the more technical and practical dimensions of coastal protection (assignment 1), another tendency can be recognized: namely, to consider only those findings that have been approved by their above-mentioned 'in house' units in charge of monitoring and assessing the ongoing scientific debate (assignment 3).

As a consequence, the dominant tendency in both the divisions in charge of monitoring and assessing new scientific and technical knowledge is to avoid direct reference to the general scientific debate on climate change as institutionalized in the framework of IPCC; this is the case as well as in the divisions of more practical and technical responsibility. The exception is in the event of 'our own experts' confirming and accepting the findings of the general debate as sufficiently certain.

Thus, a particular corpus of expertise takes shape. Being an aggregate of scientific knowledge, practical experience and administrative forehandedness, it can also be seen as a scientific–administrative hybrid. Its (expected) practical use is generally to hold off uncertainty from administrative planning and especially the demand for costly additional investments without being able to present well-grounded reasons in terms of safe knowledge. But, at the same time, it is an example of a rather paradoxical transformation of knowledge: striving for precaution without accepting uncertainty (as a core element of the precautionary principle), thus retaining the deterministic and empirically based safety discourse and the related routines of assessing and safeguarding coastal protection on the one hand, but—on the other hand—without openly rejecting the probabilistic approach of the risk discourse that has become dominant within the framework of today's climate research community.

On the level of political decision making in the domain of coastal protection within the costal area under investigation, this procedure continues so far to be accepted—even at the price of apparent conceptual inconsistencies with regard to the (more risk-orientated) concepts in other federal states (*Länder*) in charge of other parts of the rather narrow German North Sea coast (most notably Schleswig Holstein). The reasons for this might be that this does not require additional investment and that the authorities do not have to intimidate the public with bad news about expected risks or those already in existence. Another very probable reason is that, so far, no extreme events have had to be dealt with in the area since the debate on human-induced climate change has gained momentum. Consequently, the dominant safety discourse had not yet to pass a 'reality test'.

Flood Protection in German River Basins, Incorporating Risk and Uncertainty

In August 2002 heavy rains led to unprecedented floods in Central Europe and caused severe damage and the loss of 100 human lives in Austria, the Czech Republic and in south-east Germany. Compared to floods in the past (along the River Rhine in the 1990s, for example), the damage was much greater. Around 100 000 people had to be evacuated. The total economic losses due to natural disaster in that year were estimated at about 15–16 billion euros, much of it uninsured. The highest losses occurred in Germany, at nine billion euros (Becker & Grünwald, 2003). The biggest losses refer to damage to private property, with a portion of 45.6 per cent (Nachtnebel, 2003). Loss of infrastructure (18.4 per cent) included damage to railway lines, highways and regional traffic lines, electric supply networks, communication networks and pipelines. Bridges, mostly built over the last twenty years, were destroyed to a large extent. Damage from water (and sewage) brought health hazards from contamination by chemical leaks, fuel and other pollutants leaking from damaged industrial plants and enormous amounts of garbage (Wisner *et al.*, 2004, p. 203). The severe flood event was caused by a so-called 'V-b circulation', a cyclone that developed in the northern Mediterranean and travelled north-east from Genoa to the concerned countries.[7] Some areas of Saxony, the region hit hardest, experienced the heaviest rainfalls ever reported in Germany. The rainfall caused extreme floods in the basin of the

River Elbe; the floods were considered rarer than a 100-year event (an event that statistically occurs every 100 years). Others even considered them as rare as a 1000-year event. Either way, many hydrological records were broken (Mechler & Weichselgartner, 2003; Nachtnebel, 2003).

While Europe has never been exempt from floods (see above), the severity of the recent series of disasters seemed to shock not only the victims, but governments, planners and insurers as well (Wisner *et al.*, 2004, p. 201). "It was as if wealth, infrastructure, and order were being unfairly challenged by nature, in societies that considered themselves immune or robust, unlike the less developed countries" (Wisner *et al.*, 2004, p. 201). Immediately afterwards the discussion about the causes of such an extraordinary flood started. Two streams of argument can be differentiated. As both the frequency of floods and extreme rainfall events have increased, frequently raised arguments—especially the popular and mass media perception—referred to the impact of global warming and climate change (see Nachtnebel, 2003, pp. 6–8; Wisner *et al.*, 2004, p. 201). The second point of reference is direct human interventions in the river basin as an additional and worsening cause—a shift explaining flood disasters as caused by people and not just by water. The consequence of those manifold interventions became obvious. In summary: modified and intensified land use, including urbanization, sealing off large areas in the basin, river engineering works such as canalization of rivers, and losses of the retention capacity in the basin due to flood protection measures (for example dykes), increases the flood damage potential. In fact, people have settled in the historical flood plains over the past decades, leading to economic values accumulating in prone areas. Entire suburbs, for example in the city of Dresden, have been built on flood plain areas (Mechler & Weichselgartner, 2003). For the most part, these are associated with inappropriate reliance on the safety provided by flood protection measures (ZENEB, 2002).

After 2002, discussions on flooding intensified in several affected European countries, dealing with the question of how to cope with flood risk in future. At the same time awareness arose that "rivers should be allowed to flow freely within their valleys, enabling the flood plains to be restored to exactly that: flood plains" (Wisner *et al.*, 2004, p. 202). Here again, the media and popular conceptions of floods gained influence acknowledging respective "needs of nature" and drawing on inappropriate behaviour of people (Wisner *et al.*, 2004, p. 202).

In Germany the situation was special since general elections took place shortly after the flood event. The elections were in their final stage and they were perceived as a very close race. Intensive and immediate financing and assistance during the floods "boosted the government's poor standing in the polls during the run-up to the elections" (Mechler & Weichselgartner, 2003, p. 2). Such firm handling of the flood crisis was accompanied by a package of measures. The so-called Five-Point-Programme was presented at a national conference on floods after the disaster. Therewith the German government asserted to avoid such devastating floods from happening again. The programme was to "draw the lessons from the flood disasters of the last few years, rather than merely paying lip service" (BMU, 2003b).

Meanwhile (10 March 2005), these measures entered into force as the 'Flood Control Act' (BMU, 2005). Under the new Act the *Länder*[8] are *obliged* to designate more areas as flood plains than before. Water segments, where flood plains can be selected to lower the risk of flood damage, have to be identified. The *Länder* are obliged to inform the public about their decisions, and the concerned public has to be integrated into the decision-making process. For areas with a high potential of damage, flood plains have to be designated within five years, in flood-prone areas within seven years.

The *Länder* have to draw up plans co-ordinating flood protection along the rivers within four years. In the process of developing these plans, the interests of upstream and downstream users of a water body will be harmonized. The underlying insight is that every construction of flood defence built upstream may increase the risk downstream. In addition, the *Länder* have to designate flood-prone zones. This intends to raise *awareness* among the general public and the planning authorities, as more than 200 dam failures along the rivers Elbe and Mulde proved that dams and walls do not provide absolute protection against floods (BMU, 2003b, 2005). Flood plains and flood-prone zones have to be marked in spatial plans and development plans in order to point out the danger of flooding at an early stage. The basis for designating flood plains is the so-called 100-year flood (BMU, 2005). Since the Act has entered into force, planning new housing areas in flood plains is for the first time prohibited by federal law. No new buildings may be planned in these areas.[9] Another 'innovation' of the Flood Control Act is its appeal to make sure that potential damage to individuals will be as low as possible. In flood zones, computing centres and oil-fired heating systems, for example, should not be located in basements (BMU, 2003b).

The key message of the Five-Point-Programme consists of two aims. According to the slogan, "Give our rivers more space—before they take it themselves" (BMU, 2003a, 2003b), the German federal government insisted publicly on the necessity of *prevention* and *precaution*. Since flood protection was defined as an issue of *spatial planning*, thus setting up flood-risk regions as priority areas, any planning action must be compatible with this priority purpose (Friesecke, 2004, p. 10). In addition, more weight was added to *participatory* procedures. The message is 'precaution as process' (cf. Stirling, 2003). The second message is that there remains an amount of *risk*, since flood protection measures cannot guarantee absolute safety.

However, the content of the Five-Point-Programme was not novel at all. It is very similar to guidelines recommended by the Working Group on Water of the German Federal States (LAWA). These guidelines depict a forward-looking model of flood protection. Already in June 1994 the German Ministers for the Environment had instructed the LAWA to develop the guidelines, which were drawn up in November 1995 (see ZENEB, 2002). However, unlike the governmental programme of 2002, its stipulations never became mandatory regulation due to political resistance. It needed exceptional political circumstances to disprove and to de-legitimize the safety discourse, dominant also in this domain until then.

Discussion

A comparison of the two cases reveals striking similarities and differences. Options to address risks and uncertainty, although in a different way, exist in case A as well as in case B. But in case A, basic assertions of the general debate on climate change and risk are downplayed and partly transformed; in case B they have been accepted widely and transferred, thus leading to a substantial and concrete action programme.

Similarities and differences reflect particular features of the steering context. Three dimensions turn out to be relevant: different institutional settings and related actor settings within the political–administrative system (I), the occurrence or non-occurrence of 'reality tests' to be passed (II), and particular external situational factors (III).

Dimension I

Within the case of coastal protection (case A) *public administration* spans the 'boundary' between science and policy in the sense that it predominantly accepts those findings of climate research which match with the 'safety discourse'. This discourse is based on *experience* and *routines* regarding dyke safety and on a more or less linear understanding of climate change. Elements of the 'risk discourse', which in contrast underlines the potential non-linearity of further climate change and related effects such as sea-level rise, are doubted or ignored unless they can be categorized as *proven results*. Thus, risk is not addressed publicly and uncertainties are not considered publicly. In contrast, the 'risk discourse', as pursued in the IPCC-community, emphasizes prognostic uncertainties due to general deficiencies in knowledge and insufficient data availability as a starting point. Actually, in case B the federal government addressed the precautionary principle and publicly made clear that risky constellations cannot be avoided once and for all and that flood protection measures guarantee only limited safety.

In case A, a particular administrative facility responsible for assessing findings of the general scientific debate on climate change (with relevance to coastal protection) has proven to be the key actor. Here, administrative actors in charge have successfully assumed the authority to make political decision makers accept a perspective focused on the safety discourse. In fact, there is no institutional equivalent in case B. Here, the pivotal role of political (governmental) actors is evident. This might be the very reason why in this case it proved possible to quickly institutionalize the risk discourse and to oblige the administration responsible for flood protection to adopt it. However, this is only one out of several explaining factors.

Dimension II

The case of coastal protection on the North Sea coast, whilst staying within the boundaries of a deterministic and more or less linear perspective, can be explained by the fact that up to now—fortunately—the protective systems in the area did not have to pass a 'reality test'.

As for case B, in contrast, the severe riverine floods of 2002 can be interpreted as such a reality test, disproving the idea that it is possible to calculate and to ensure 'safety' under all realistic circumstances. Comparing both cases provides strong arguments that real disasters, such as the extreme floods in Germany in August 2002, may help to undermine the so far dominant safety discourse and to establish elements of a risk discourse, demonstrating that risks are real and that ignoring them can lead to severe damage—not only in the future but even today—and that extreme events may occur anew even tomorrow. The "collapse of confidence in engineered flood protection" (Wisner *et al.*, 2004, p. 203) fostered a paradigm shift towards 'living with floods'—an idea to be considered normal, which can be and must be accepted.

Table 1. Dimensions of the steering context

	Institutional level/actors	Reality test	Particular political framings
Case A	German *Länder*: Public administration	No	Stable
Case B	Federal government: 'red–green' coalition	Severe floods	Elections

This shift includes the opinion that rivers, their banks and their flood plains provide valuable 'ecological services' (which can include the absorption of some flood water). This encompasses a growing acceptance of the need to understand the function of rivers and their flow regimes in relation to the wider environment. Following the precautionary principle, adaptive measures have to be considered and decided upon today. Thus, the flood-protection issue is not framed any longer as a mere engineering challenge, but as a task for and responsibility of *spatial planning*. In fact, the risk discourse no longer seems to be questioned effectively publicly.

Dimension III

A further decisive factor in case B derived from the approaching elections. There can be no doubt that the coincidence of the damages produced by the floods in August 2002 and the upcoming elections as a specific challenge to the political class to demonstrate commitment worked very much in favour of accepting the risk perspective at the political level. In fact, the combination of *spatial planning* and precautionary principle fitted very well in the discourse of 'ecological modernization', pursued by the government in charge during that period, a coalition of Social Democrats and the Green Party. Thus, the Five-Point-Programme was accepted without any major problems.

Summary

Table 1 summarizes the configuration discussed so far. Policy change has proven to be no linear problem-solving process. Rather, one witnesses a dependency on different contextual factors. This is in line with Kingdon's (1984) policy window-approach. In case B, in which policy change occurs, Kingdon's problem stream refers to the flood disaster in 2002. The equivalent of the policy stream is represented by the LAWA guidelines of 1995, as an innovative concept available immediately after the extreme event. Public opinion, the elections and—at that time—the lack of discursive opposition represent Kingdon's politics stream. The policy window consists of a coincidence of a publicly perceived problem, that is, an event in the politics stream, and political actors willing to implement a risk discourse.

Conclusion

Climate change can be explained only as a result of many different elements interacting with each other. Thus, the knowledge so far available remains limited and bears fundamental uncertainties in prognosis. More sophisticated models may reduce uncertainty. However, with regard to climate change and its potential threats to society there is no way to substantially avoid or even overcome uncertainty.

It goes without saying that political decision makers and administrative officers, for example those in charge of coastal protection and (riverine) flood, feel uneasy with such a constraint. But in order to develop and implement effective strategies they have to deal with uncertainty anyway.

Doing so, they can pursue different strategies. They can, for example, reject assertions of climate research as insufficiently confirmed by empirical evidence. Referring to their particular knowledge and skills, they can even be successful in making political decision makers adopt this perspective. But equally they can be compelled by political coercion to adopt an uncertainty-perspective to substantially reset protective schemes in force so far.

Accepting or rejecting uncertainty as a core message of climate research means transferring or transforming respective knowledge as a part of 'boundary work', thus maintaining or developing different perceptions of risk and related risk cultures. Here, institutional settings and professional cultures seem to be of particular relevance.

Differing risk perceptions are far from being unequivocal and undisputed. Rather, they should be seen as competing discourses striving for dominance within a discourse arena. This applies likewise to the 'safety discourse' and the 'risk discourse' outlined above. Which of them successfully prevails does not necessarily mirror the quality of its particular arguments. Actually, different settings of arguments prove relevant and convincing for different categories of actors being bound to different frameworks of interest and rationalities (especially scientists vs. political decision makers vs. administrative officers). Moreover, which of the competing discourses wins the day depends largely on whether and what kind of window of opportunity is open or not. In other words, which of the competing discourses will prevail can hardly be foreseen.

Nevertheless, the outcome can be influenced substantially by societal actors. Two aspects seem to be of paramount importance. The first of these is preparedness on a conceptual level in order to present a particular option at the very moment when the window of opportunity starts opening; this particular option must have been developed in advance and held ready. Secondly, the actors must have at their disposal means to immediately propose decisions when the window of opportunity opens. Here, although dependent in many respects on other societal actors, the state, as formally legitimized (by constitution) and having the legal means and the power to enact its option, continues to be a key actor.

There is some empirical evidence that once such a policy change has taken place and subsequent concepts have started to be enacted, the effect is not confined to the domain of the individual case to which it refers. As a matter of transversal shifts within the broader arena of public discourse, changes brought about in particular areas may also contribute to making changes easier in other sectors of the arena that are similarly dealing with controversies concerning how to cope with the scientific debate on climate change. This seems to apply not least to the fields of riverine flood protection on the one hand and to coastal protection on the other. In fact, five years after the flood events that hit the German midland in 2002 the former stiff reluctance against accepting uncertainty and risk as core issues of today's climate research is also losing ground on the German North Sea coast. This can be explained against the background of a conjunction of heterogeneous factors: the general rise of the debate on climate change (in particular referring to above average temperatures during the last few years in Central Europe and as a response to the 4th Assessment Report of the IPCC) and significant changes in the political–administrative system of Bremen (personal replacement due to the regular retirement of a leading coastal protection officer on the one hand and the recent formation of a new government in Bremen based on a 'red–green' coalition replacing the more conservative former 'red–black' coalition on the other hand).

If similar shifts could be confirmed in comparable constellations, this would mean that paradigm shifts (from 'safety discourse' to 'risk discourse') do not necessarily presuppose 'reality tests' in the sense of experiencing damage and harm produced by unexpected extreme events. In other words, it could be possible to prepare for and adapt to the occurrence of novel extreme events as long as there is still enough time left to avoid disasters by thorough and comprehensive precaution.

Notes

1. For further information relating to the two cases see www.krim.uni-bremen.de and www.innig.uni-bremen.de.
2. Their work is based on their experience in the context of the FAO and refers to the global nitrogen cycle.
3. The empirical basis of the finding consists of 50 interviews conducted in 2003 and 2006. They included officers in charge of different subtasks in the realm of coastal and flood protection at all levels (local, district, federal state) and the involved political entities in terms of geography: the federal states, so-called *Länder*, Niedersachsen, Bremen, Hamburg. In addition, political actors (parties represented in parliaments, NGOs) were interviewed as well as further non-state actors, such as dyke associations. The interviewees—and further analysed documents— were selected in a way that ensures a complete coverage of the political–administrative system of coastal and flood protection in the area. In analysing these sources, Keller (1998, 2004) was referred to.
4. http://www.artec.uni-bremen.de/files/projekte/Endbericht_KRIM.pdf; subsequently quoted as Lange *et al.* (2005). The area under investigation comprises the coastline of the mainland around the Jade Bay and the respective administrative districts (*Landkreise*) Friesland, Wesermarsch and Cuxhaven, including the island of Wangerooge and the estuary of the River Weser up to the city of Bremen. Large parts of the area are situated at up to 2 m below the mean high tide margin. Because of this topology and because it consists largely of supple marine sediments, the coastal area is prone to erosion and flooding.
5. Coastal protection in Germany is planned and executed at three administrative levels (nation state, federal states, local/municipality level) covering four different federal states—each of them following more or less different planning concepts including different forms of providing information and allotting responsibility to citizens living in flood-prone areas (Lange *et al.*, 2005, pp. 22 ff.).
6. In detail, the procedure is actually more sophisticated (see Lange *et al.*, 2005, p. 31.)
7. In the last decade similar circulation patterns caused the Odra flood in 1997 and the Vistula flood in 2001 (Nachtnebel, 2003).
8. See note 5.
9. Exceptions are possible if nine closely defined requirements are met. All of them have to be fulfilled completely in every individual case. They include that the municipality concerned has no alternative for human settlement development, that no lives are at risk and no significant property damage is to be expected, and that the structure of new buildings is adapted to flood events (BMU, 2005).

References

Arts, B., van Tatenhove, J. & Leroy, P. (2000) Policy Arrangements, in: Jan van Tatenhove, Bas Arts & Pieter Leroy (Eds) *Political Modernisation and the environment. The Renewal of Environmental Policy Arrangements*, pp. 53–70 (Dordrecht: Kluwer).
Bäckstrand, K. & Lövbrand, E. (2006) Planting Trees to Mitigate Climate Change: Contested Discourses of Ecological Modernization, Green Governmentality and Civic Environmentalism, *Global Environmental Politics*, 6, pp. 51–75.

Beck, U. (1992) *Risk Society. Towards a New Modernity* (London: Sage).

Becker, A. & Grünwald, U. (2003) Flood risk in Central Europe. *Science*, 300, p. 1099.

BMU (Federal Ministry for the Environment, Nature Conservation and Nuclear Safety) (2003a) Jürgen Trittin: effective climate protection is flood prevention for tomorrow. European experts meet to discuss prevention measures, Press statement No. 012, 5 February 2003.

BMU (2003b) Trittin presents draft Flood Control Act. Give our rivers more room—before they take it themselves, Press statement No. 143, 8 August 2003.

BMU (2005) New Flood Control Act enters into force. Preventive flood protection is improved significantly, Press statement No. 111, 9 May 2005.

Bonß, W. (2004) Zwischen Verwendung und Verwissenschaftlichung. Oder: Gibt es eine 'Lerngeschichte' der Politikberatung?, *Zeitschrift für Sozialreform*, 1–2, pp. 32–45.

Dryzek, J. S. (1997) *The Politics of the Earth. Environmental Discourses* (Oxford: Oxford University Press).

Eser, U. (2001) Die Grenze zwischen Wissenschaft und Gesellschaft neu definieren: Boundary work am Beispiel des Biodiversitätsbegriffs, *Verhandlungen zur Geschichte und Theorie der Biologie*, 7, pp. 135–152 (Berlin: VVB).

European Environment Agency (EEA) (2004) *Impacts of Europe's changing climate. An indicator-based assessment* (Summary). http://repoits.eea.europa.eii/climate_report_2_2004/en/impacts_of_europes_changing_climate.pdf (30 September 2007).

Friesecke, F. (2004) Precautionary and Sustainable Flood Protection in Germany—Strategies and Instruments of Spatial Planning. Paper presented at the 3rd FIG Regional Conference, Jakarta, Indonesia, 3–4 October.

Grundmann, R. (1999) *Transnationale Umweltpolitik zum Schutz der Ozonschicht. USA und Deutschland im Vergleich* (Frankfurt: Campus).

Grunwald, A. (2008) Steering for Sustainable Development facing uncertainty and incompleteness of knowledge, in: J. Newig, J.-P. Voß, J. Monstadt (Eds) Governance for Sustainable Development: Steering in Contexts of Ambivalence, Uncertainty and Distributed Power. Routledge: London (in press)

Gyrien, T. F. (1983) Boundary-work and the demarcation of science from non-science: strains and interests in professional ideologies of scientists, *American Sociological Review*, 48, pp. 781–795.

Gyrien, T. F. (1995) Boundaries of science, in: S. Jasanoff *et al.* (Eds) *Handbook of science and technology studies*, pp. 393–443 (Thousand Oaks, Calif.: Sage Publications).

Haas, P. A. (1992) Introduction. Epistemic Communities and International Coordination, *International Organization*, 46, pp. 1–35.

Hajer, M. A. (1995) *The Politics of environmental Discourse. Ecological Modernization and the Policy Process* (Oxford: Clarendon Press).

Hirsch-Hadorn, Gertrud, Kissling-Näf, Ingrid & Pohl, Christian (2004) How to design Interfaces between Science and society: Lessons From Platforms for Knowledge Communication in Switzerland, in: Frank Biermann, Sabine Campe & Klaus Jacob (Eds) *Proceedings of the 2002 Conference on the Human Dimensions of Global Environmental Change "Knowledge for the Sustainability Transition. The Challenge for Social Sciences"*, pp. 285–291 (Amsterdam, Berlin, Potsdam and Oldenburg: Global Governance Project).

IPCC (2001) Climate *Change (2001) Impacts, Adaptation, and Vulnerability.* Available at http://www.ipcc.ch/pub/wg2SPMfinal.pdf (30 September 2007).

IPCC (2007) 4th Assessment Report: *Summary for policy* makers. Available at http://www.ipcc.ch/WG1_SPM_17Apr07.pdf (30 September 2007).

Jasanoff, S., Markle, G. E., Petersen, J. C., Pinch, T. (Eds) (1995) Handbook of science and technology studies (Thousand Oaks, Calif.: Sage Publications).

Keller, R. (1998) *Müll – die gesellschaftliche Konstruktion des Wertvollen – die öffentliche Diskussion über Abfall in Deutschland und Frankreich* (Opladen: Westdeutscher Verlag).

Keller, R. (2004) *Diskursforschung: Eine Einführung für SozialwissenschaftlerInnen* (Opladen: Leske und Budrich).

Kingdon, J. W. (1984) *Agendas, alternatives, and public policies* (Boston: Little, Brown and Co.).

Lane, R. E. (1966) The Decline of Politics and Ideology in a Knowledgeable Society, *American Sociology Review*, 31, pp. 649–662.

Lange, H., Haarmann, M., Wiesner-Steiner, A., Voosen, E. (2005) Klimawandel und präventives Risiko- und Küstenschutzmanagement an der deutschen Nordśeekuste (KRIM) - Teilprojekt IV: Politisch-administrative Steuerungsprozesse (PAS). Bremen. Available at http://www.krim.uni-bremen.de/endberichte/endbericht_tp4.pdf (30 September 2007)

Mechler, R., Weichelgartner, J. (2003) Experiences with financing risk management and reconstruction in the floods of 2002 in Germany. Presentation held on the occasion of the Third Annual IIASA-DPRI

Meeting on Integrated Disaster Risk Management: Coping with Regional Vulnerability Kyoto International Conference Hall, June 29 - July 7, 2003, Kyoto

Nachtnebel, H.P. (2003) New Strategies for Flood Risk Management after the Catastrophic Flood in 2002 in Europe. In: Disaster Prevention Research Institute (DPRI) of the Kyoto University and International Institute of Applied Systems Analysis(IIASA), Laxenburg, Austria: Third DPRI-IIASA International Symposium on Integrated Disaster Risk Management (IDRM-2003), 3–5 July, 2003, Kyoto International Conference Hall, Kyoto, Japan.

Norse, D. & Tschirley, J. B. (2000) Links between science and policy making, *Agriculture, Ecosystems and Environment*, 82, pp. 15–26.

Nullmeier, F. (1993) Wissen und Policy-Forschung. Wissenspolitologie und rhetorisch-dialektisches Handlungsmodell, in: A. Héritier (Ed.) *Policy Analyse. Kritik und Neuorientierung*, pp. 175–198 (Opladen: Westdeutscher Verlag).

Paavola, J. & Adger, N. W. (2004) Knowledge or Participation for Sustainability? Science and Justice in Adaptation to Climate Change, in: Frank Biermann, Sabine Campe & Klaus Jacob (Eds) *Proceedings of the 2002 Conference on the Human Dimensions of Global Environmental Change "Knowledge for the Sustainability Transition. The Challenge for Social Sciences"*, pp. 175–183 (Amsterdam, Berlin, Potsdam and Oldenburg: Global Governance Project).

Potthast, T. (1999) *Die Evolution und der Naturschutz: Zum Verhältnis von Evolutionsbiologie, Ökologie und Naturethik* (Frankfurt am Main: Campus).

Scharpf, F. W. (2000) *Interaktionsformen Akteurszentrierter Institutionalismus in der Politikforschung* (Opladen: Leske und Budrich).

Schiller, Frank (2005) *Diskurs über Nachhaltigkeit – Zur Dematerialisierung in den industrialisierten Demokratien* (München: Ökom-Verlag).

Stehr, N. (1994) *Arbeit, Eigentum und Wissen. Zur Theorie von Wissensgesellschaften* (Frankfurt am Main.: Suhrkamp).

Stirling, A. (2003) Risk, uncertainty and precaution: some instrumental implications from the social sciences, in: Frans Berkhout, Melissa Leach & Ian Scoones (Eds) *Negotiating Environmental Change*, pp. 33–76 (Cheltenham, UK: Edward Elgar).

Szarka, J. (2004) Wind Power, Discourse Coalitions and Climate Change: Breaking the Stalemate?, *European Environment*, 14, pp. 317–330.

Takacs, D. (1996) *The Idea of Biodiversity. Philosophy of Paradise* (The John Hopkins University Press: Baltimore [u.a.]).

Viehöver, W. (1997) 'Ozone thieves' and 'hot house paradise'. Epistemic communities as cultural entrepreneurs and the reenchantment of the sublunar space. PhD thesis, European University Institute, Florence.

Voß, J.-P., Newig, J., Kastens, B., Monstadt, J. (2008) Steering for Sustainable Development – A typology of empirical contexts and theories based on ambivalence, uncertainty and distributed power, in: J. Newig, J.-P. Voß, J. Monstadt (Eds) Governance for Sustainable Development: Steering in Contexts of Ambivalence, Uncertainty and Distributed Power. Routledge: London (in press)

Weingart, P. (2001) *Die Stunde der Wahrheit – vom Verhältnis der Wissenschaft zu Politik, Wirtschaft und Medien in der Wissensgesellschaft* (Weilerswist: Velbrück Wissenschaft).

Weingart, P., Engels, A. & Pansegau, P. (2002) *Von der Hypothese zur Katastrophe. Der anthropogene Klimawandel im Diskurs zwischen Wissenschaft, Politik und Massenmedien* (Opladen: Leske + Budrich).

Wisner, Ben, Blaikie, Piers, Cannon, Terry & Davis, Ian (2004) *At risk. Natural hazards, people's vulnerability and disasters* (London and New York: Routledge).

ZENEB (Zentrum für Naturrisiken und Entwicklung Bonn/Bayreuth) (2002) Zusatzbeitrag. Floods in Europe: Lessons learned?. Available at http://www.giub.uni-bonn.de/zeneb/akt/bericht_wvr.pdf (31 December 2006).

Managing Uncertainties in the Transition Towards Sustainability: Cases of Emerging Energy Technologies in the Netherlands

INEKE MEIJER & M. P. HEKKERT

Introduction

This article focuses on a particular aspect of sustainable development, namely sustainable technology development. The relation between technology and sustainability is complex and paradoxical (Grubler, 1998). Apart from the advantage of creating economic growth and societal benefits, current use of technologies may cause severe environmental problems, such as pollution and depletion of resources. However, technologies may also lead to a more efficient use of resources, less stress on the environment, and cleaning of the environment. Thus, steering technological change towards sustainability—also referred to as "sustainable technology development" (Weaver *et al.*, 2000)—is a fundamental part of governance for sustainable development.

This article does not refer to technology development in the narrow sense, but to the development of technology in interaction with the socio-institutional system in which the technology is embedded (Hekkert *et al.*, 2006). Creating technological change aimed at sustainable development does not only involve technological change but also changes in the social and cultural dimension, such as user practices, regulation and industrial networks (Elzen & Wieczorek, 2005; Geels, 2002b) (see also Grunwald (2007), an article in this special issue). The increasing recognition among scientists and policy makers of this system level of change has led to a rapid diffusion of concepts such as transitions or industrial or socio-technological transformation (Elzen & Wieczorek, 2005; Elzen *et al.*, 2004; Geels, 2002b; Sagar & Holdren, 2002; Smith *et al.*, 2005). A transition, as referred to in this article, is defined as a major, long-term socio-technological change in the way societal functions (such as the supply of energy) are being fulfilled (Geels, 2002b). This long-term transformation at the level of society as a whole, in turn, consists of a sequence of short-term innovations (Geels, 2002b).

The central idea of this special issue is that steering sustainable development is problematic due to the ambivalence of goals, the uncertainty of knowledge about system dynamics, and the distributed power to shape system development (Geels, 2002a; Grubler *et al.*, 1999; Jacobsson & Johnson, 2000; Kemp & Soete, 1992; Unruh, 2002; Voß, Newig, Karstens, Monstadt, Noltring, 2007). First, the systemic character of transitions implies that a wide diversity of actors is involved and that none of the actors can achieve a transition alone (distributed power) (Meijer *et al.*, 2006; Smits & Kuhlmann, 2004). In other words, the behavior of various actors collectively determines the speed and direction of a transition. Because actor-behavior has such an important influence on the overall transition, this article aims to deepen our understanding of transitions by applying an actor-perspective. Second, due to the large diversity of actors, there are several perceptions of the final objective of the transition and these perceptions constantly change during the transformation process, in order to adapt to new situations (ambivalence of sustainability goals) (Rotmans *et al.*, 2001). In order to reach each of the possible objectives, many separate innovation decisions—decisions to develop or adopt a technological innovation that will contribute to the goal of the overall transition—have to be taken (Suurs *et al.*, 2004). Thus, the ambivalence of sustainability goals leads to an infinite number of possible outcomes and an infinite number of possible innovation decisions. Third, due to the long time span of transitions and the co-evolution of technological and societal changes, knowledge of the dynamics of innovation systems is limited and uncertainty about the possible effects of innovation activities is high (uncertainty of knowledge about system dynamics; see also Grunwald (2007) and Lange & Garrelts (2007) in this special issue). As a result, the actors involved in transition processes perceive great uncertainty, both about the final outcome of a transition and about each of the short-term innovation decisions. The uncertainties perceived by the actors play an important role, since they influence the actors' decisions and behavior greatly. Perceived uncertainties can be regarded as positive when they stimulate actors to engage in novel sustainable technological trajectories. However, uncertainties may also have the effect that actors do not dare to invest in these desired directions of change. In that case uncertainties block actors to undertake activities that are essential for achieving a transition towards sustainability. Since the behavior of individual actors collectively determines the speed and direction of a transition, these perceived uncertainties are likely to influence the transition as a whole. In order to improve our knowledge of the underlying system dynamics of transitions towards sustainability, insight into the various types of perceived uncertainties and actors' responses to these uncertainties is a prerequisite.

The aim of this article is to come to a better understanding of the influence of perceived uncertainties on the innovation decisions of actors involved in transition processes, in order to contribute to the difficult task of steering transitions towards sustainability. Ideally, governmental policy should contribute to the management of uncertainties, aiming at stimulating transitions. In some cases this implies that uncertainties need to be created in order to stimulate new directions of change, while in other cases governmental policy needs to be adapted when it is considered an important source of uncertainty that hamper transitions (Meijer *et al.*, 2006; 2007). This article aims to demonstrate which types of perceived uncertainties influence the innovation decisions of actors involved in sustainable technology development and how steering initiatives of the Dutch government influence these uncertainties, by examining two empirical cases of emerging sustainable energy technologies in the Netherlands. The first case concerns the development of micro-CHP (Combined generation of Heat and Power at domestic scale). The second case focuses on the introduction of biofuels (liquid fuels produced from biomass and used in the transport sector). In the Netherlands, expectations of the potential contribution of both micro-CHP and biofuels to the transition towards sustainability are high and many initiatives are currently being developed. However, since these technologies are still in an early stage of development, uncertainties about the future are likely to be high. That makes micro-CHP and biofuels interesting cases to study the role of uncertainties in emerging sustainable energy technologies. The main question of this article is 'Which types of uncertainties are perceived by the actors involved in emerging technological trajectories and how do they deal with these uncertainties?'.

Special attention is paid to the perspective of the entrepreneurs (the business firms involved in developing and implementing the new technology), since technological innovation cannot take place without entrepreneurs who dare to take action. In order to relate this entrepreneurial perspective to the governmental perspective, the article also focuses on how steering initiatives of the Dutch government influence the perceptions and behavior of the entrepreneurs.

Uncertainty and Transition

Transitions involve a wide variety of actors, each playing their own role in the transformation process. The most important role in technological innovation and transition processes is the role of the entrepreneur who turns the potential of new knowledge, networks and markets into concrete actions to generate—and take advantage of—new business opportunities (Hekkert *et al.*, 2006). Different types of actors can perform the role of entrepreneur. Entrepreneurs can be technology developers wanting to market their technology, but they can also be adopters (buyers and users of the technology) who seek profit in the application of the technology.[1] Furthermore, entrepreneurs can be new entrants with visions of business opportunities in new markets, or established companies who diversify their business strategy to take advantage of new developments (Hekkert *et al.*, 2006). Because of differences in objectives, resources and so on, one expects that different types of actors will have different perceptions of uncertainties and that they will react differently to perceived uncertainties. This section describes the theoretical framework used to study the different perceptions of uncertainties and the reactions of the entrepreneurs to these uncertainties in the empirical cases. This framework was based on a review of uncertainty and innovation literature, as reported in previous articles (Meijer *et al.*, 2006; 2007).

Sources of Perceived Uncertainty

Uncertainty arises when the actors involved in a transition do not know what the effects of their innovation decisions will be. In this article, the term 'uncertainty' is defined broadly as "any deviation from the unachievable ideal of completely deterministic knowledge of the relevant system" (Walker *et al.*, 2003). It is important to note that gathering information cannot always reduce uncertainty. Uncertainty can exist even in situations where much information is available (Koppenjan & Klijn, 2004; Van Asselt, 2000).

An ongoing uncertainty debate among scholars is the discussion about objective versus perceived uncertainty (e.g. Jauch & Kraft, 1986; Kreiser & Marino, 2002). Supporters of the objective view of uncertainty define uncertainty as a characteristic of the environment that can be measured objectively (Dess & Beard, 1984). Supporters of the perceptive view of uncertainty argue that uncertainty depends on the individual and cannot be measured objectively (Milliken, 1987). The term 'perception' refers to the process by which individuals organize and evaluate stimuli from the environment. The existence of information itself lacks meaning until an individual perceives it (Corrêa, 1994). Environments are, therefore, neither certain nor uncertain but are simply perceived differently by different actors. In this article, we are interested in the innovation behavior of actors. The intention is to analyze if uncertainties stimulate or block actors fulfilling certain key activities essential for achieving a transition. Since perceived uncertainties, and not objective uncertainties, influence this behavior, the focus is on 'perceived uncertainties'. In the remainder of this article, 'uncertainty' refers to 'perceived uncertainty'.

This article classifies perceived uncertainties according to each of their sources. The source of uncertainty is the domain of the (organizational) environment which the decision maker is uncertain about (Milliken, 1987). Distinguishing different sources is important for choosing appropriate strategies to cope with the uncertainty (Wernerfelt & Karnani, 1987). Based on an extensive literature review, reported in Meijer *et al.* (2006), the following set of uncertainty sources are proposed with respect to innovation decisions of entrepreneurs.

(i) *Technological uncertainty.* This source includes uncertainty about the characteristics of the new technology (such as costs or performance), uncertainty about the relation between the new technology and the infrastructure in which the technology is embedded (uncertainty to what extent adaptations to the infrastructure are needed) and uncertainty about the possibility of choosing alternative (future) technological options. Uncertainty about the direction of the transition process is reflected in this source of uncertainty.

(ii) *Resource uncertainty.* This source includes both uncertainty about the amount and availability of raw material, human and financial resources needed for the innovation, and uncertainty about how to organize the innovation process (e.g. in-house or external R&D, technology transfer, education of personnel). Resource uncertainty resides both at the level of the individual firm, as well as at the level of the innovation system.

(iii) *Competitive uncertainty.* Whereas technological uncertainty includes uncertainty about competing technological options, competitive uncertainty relates to uncertainty about the behavior of (potential or actual) competitors and the effects of this behavior.

(iv) *Supplier uncertainty.* Uncertainty about the actions of suppliers amounts to uncertainty about timing, quality and price of the delivery. Supplier uncertainty becomes increasingly important when the dependence on a supplier is high.

(v) *Consumer uncertainty.* Uncertainty about consumers relates to uncertainty about consumers' preferences with respect to the new technology, uncertainty about the compatibility of the new technology with consumers' characteristics,[2] and, in general, uncertainty about the long-term development of the demand over time.

(vi) *Political uncertainty.* Political uncertainty comprises uncertainty about governmental behavior, regimes and policies. Not only changes in policy, but also ambiguity in interpretation of current policy or a lack of policy can lead to uncertainty. Another important cause for political uncertainty is unpredictability of governmental behavior. This source of uncertainty also reflects uncertainties related to the direction of transition processes. More specifically it relates to the uncertainty with which transition directions are backed by government actions and support.

Functioning of Innovation Systems

Actors can respond to perceived uncertainties in many different ways. One of the standard responses to perceived uncertainties is to delay or even abandon (innovation) decisions (Koppenjan & Klijn, 2004). In other words, perceived uncertainties might prevent actors from fulfilling certain key activities essential for achieving a transition and, thereby, they might hamper the transition as a whole (Jacobsson & Bergek, 2004). However, perceived uncertainties do not necessarily have to hinder transitions. Some scholars argue that organizations in an uncertain environment tend to be more proactive and innovative and tend to embrace more risks (Jauch & Kraft, 1986). Instead of abandoning or delaying innovation decisions, actors can also accept that innovation is inherently uncertain and consciously deal with these uncertainties. For example, if an entrepreneur perceives high technological uncertainty about an innovative technology, the entrepreneur can either decide to abandon investments or to experiment in order to learn about the new technology and, thereby, reduce uncertainty. Thus, perceived uncertainties might also induce actors to fulfill activities that contribute to the overall transition.

Although each technological trajectory is unique with respect to the technological and institutional setting that influence the transformation process (the so-called 'innovation system'),[3] recent innovation scholars have formulated a generic list of key activities that are essential for achieving a transition. Since these key activities have the function of contributing to the goal of the innovation system, which is the generation, diffusion and utilization of innovations, the term 'functions of innovation systems' (in short 'system functions') is used to describe the set of key activities[4] (Foxon *et al.*, 2005; Hekkert *et al.*, 2006; Huang & Wu, 2007; Jacobsson & Bergek, 2004; Johnson, 2001; Smith *et al.*, 2005). The following system functions are distinguished (Hekkert *et al.*, 2006):

(i)   *Entrepreneurial activities.* Experimenting by entrepreneurs is necessary to collect more knowledge about the functioning of the technology under different circumstances and to evaluate reactions of consumers, government, suppliers and competitors.

(ii) *Knowledge development.* R&D and knowledge development are prerequisites for innovation. This function encompasses 'learning by searching' and 'learning by doing'.

(iii) *Knowledge diffusion through networks.* The exchange of information through networks of actors (research institutes, governmental agencies, consumers, entrepreneurs) contributes to 'learning by interacting' and, in the case of user-producer networks, 'learning by using'. This function is especially important when a heterogeneous set of actors is involved in the innovation process.

(iv) *Guidance of the search.* Since resources tend to be limited, it is important that specific foci are chosen for further investments when various technological options exist. Without this selection, there will be insufficient resources for the individual options. This function includes those activities that can positively influence the visibility and clarity of specific needs among technology users.

(v)   *Market formation.* New technologies often have difficulty competing with embedded technologies. Therefore, it is important to facilitate the formation of markets, e.g. by the formation of niche markets or by favorable tax regimes.

(vi) *Resources mobilization.* The allocation of sufficient resources, both human and financial, is necessary as a basic input to all the activities of the innovation process.

(vii) *Creation of legitimacy/counteract the resistance to change.* In order to develop well, new technologies often have to become part of an established regime or even have to overthrow it. Parties with vested interest often oppose to this force of 'creative destruction'. In that case, advocacy coalitions (Sabatier, 1988; 1998) can create legitimacy for the new technology by putting the new technology on the agenda and lobbying for resources and favorable tax regimes.

According to this functional approach to innovation system policy, stimulating transitions implies stimulating the fulfillment of the aforementioned functions (Hekkert *et al.*, 2006; Jacobsson, 2005; Jacobsson & Bergek, 2004). Following Jacobsson & Bergek (2004), added to this is the fact that uncertainties can be an underlying force with a major influence on the functional pattern of innovation systems. Jacobsson & Bergek (2004) argued that high uncertainty in terms of technology, consumers and changing policy has blocked system fulfillment in the transition to renewable energy technologies. This would imply that policy aimed at stimulating transitions towards sustainability should also focus on the management of uncertainties to promote the fulfillment of system functions. In other words, if perceived uncertainties block the fulfillment of system functions, reducing the size of uncertainties or helping actors to cope with uncertainties may very well be a (indirect) way to stimulate desired functional patterns. The following empirical cases aim to shed light on this issue by analyzing which types of uncertainties are perceived by the actors and how they respond to these uncertainties in terms of the fulfillment of system functions.

The Case of Micro-CHP

The first empirical case focuses on the introduction of micro-CHP in the Netherlands. Combined generation of heat and power (CHP), also known as cogeneration, means that heat and power are generated simultaneously. Up to now, CHP plants have been large-scale units used for industrial processes and district heating. Currently, progress is made to apply CHP at a domestic scale (i.e. with an electrical power below 5 kW_e). This domestic application is called micro-CHP and is supposed to be a substitute for the high-efficiency boiler. The utilization of micro-CHP can lead to substantial energy savings and carbon emission reduction, since the overall efficiency is higher compared to generating space heating, hot water and electricity separately. In addition, because of the decentralized generation of electricity, distribution loss can be avoided. Therefore, in the Netherlands, micro-CHP is considered one of the promising technologies able to contribute to the transition towards a sustainable energy system (Ministerie van EZ, 2004; 2006).

Below are described those uncertainty sources that are perceived by the various actors as dominant, and whether or not these perceived uncertainties hamper the actors in fulfilling system functions. This is based largely on Meijer *et al.* (2007). The data for this case were collected by studying grey literature and conducting interviews with the main actors involved in the development of micro-CHP. In order to have a good representation of the actor groups involved, an equal number of technology developers[5] and potential adopters (i.e. potential buyers and users of the technology, such as energy companies or housing organizations[6]) were interviewed. A spokesperson of the Ministry of Economic Affairs, who is concerned with the energy (transition) policy, was also interviewed, in addition to a spokesperson of an intermediary organization that plays an important role in diffusing knowledge and lobbying for CHP in the Netherlands. The interviews took place in the summer of 2004.

Perceived Uncertainties

According to the interviewees, technological and political uncertainty appeared to be the most dominant uncertainty sources, followed by consumer uncertainty.[7] Uncertainty about resources, competitors and suppliers played only a modest role in the micro-CHP case. Below, the dominant uncertainty sources are described in more detail.

Technological uncertainty.   Since micro-CHP is not yet a 'proven technology', the most important element of technological uncertainty was uncertainty about the technology itself (uncertainty about the future performance of the micro-CHP systems in terms of reliability, investment costs, energy efficiency and so on). This uncertainty was perceived to be equally important by all the actors. Uncertainty about the relation between micro-CHP and the technological infrastructure and uncertainty about alternative (future) technological options were perceived as less important. Most of the interviewees did not foresee major technological difficulties connecting micro-CHP to the electricity grid. Nevertheless, as described below, the connection to the grid does lead to substantial political uncertainties. The interviewees indicated that they did not perceive uncertainty about the choice between different technological options, since they keep several options open or believe that each technology can occupy its own niche market.

Political uncertainty. Many interviewees experienced uncertainty about the reliability of the government in general. This lack of faith is due mainly to unexpected changes in governmental policy. Several interviewees declared that such unexpected changes can have fatal consequences for emerging technologies by pointing out the example of the sudden ending of many subsidy schemes for renewable energy by the Dutch government.

Apart from this general form of political uncertainty, the interviewees enumerated several specific policy issues that have created uncertainty in the development of micro-CHP, such as uncertainty about subsidies, energy saving norms and legal admission of individual micro-CHP owners to the electricity grid. Of special importance to the development of micro-CHP is the uncertainty about the energy taxes and electricity feed-in policy, which strongly influence the economic feasibility of micro-CHP. At the time of the interviews, it was still unclear how the application of micro-CHP would be incorporated in the energy regulation framework.

Consumer uncertainty. With respect to consumer uncertainty, one can see a clear distinction between the actors. Technology developers, on the one hand, seemed convinced about the emergence of a market for micro-CHP and believed it to be only a matter of time. Most of them claimed that uncertainties about the preferences or characteristics of consumers were small and could be reduced by market studies or pilot projects. Only one technology developer indicated that there were still major uncertainties about the market for micro-CHP. What is striking, though, is that this technology developer indicated that he/she would focus only on technological and political uncertainty and would simply ignore consumer uncertainty until the micro-CHP system was ready for market introduction.

While the technology developers seemed to have high expectations about the market for micro-CHP, the other actors (the potential adopters, the government and the intermediary organization) were more reserved. These actors did perceive uncertainty about the development of a market, for instance how large this market will be and how fast it will emerge. Two potential adopters even considered consumer uncertainty as the most important uncertainty source. They both explained that if consumers do not want micro-CHP, this will bring a stop to the entire development process.

Uncertainties in relation to system functions. Whether or not the perceived uncertainties have hindered the fulfillment of system functions depended on the type of actor in the micro-CHP case. Below, the reactions of the two main market parties are compared: the technology developers versus the potential adopters. Subsequently, the various initiatives of the Dutch government to reduce the uncertainties perceived by the market parties are discussed.

Technology developers versus potential adopters. Technology developers consciously tried to deal with the perceived uncertainties. Their activities clearly focused on the uncertainty sources that they perceived as most important, namely technological and political uncertainty. In reaction to technological uncertainty, they all initiated R&D activities. These activities seemed successful, since progress has been made in terms of the performance of the micro-CHP systems. Thus, technological uncertainty incited these actors to develop activities that contribute to the function of 'knowledge development' (function (ii) in the section 'Functioning of Innovation Systems').

In reaction to perceived political uncertainty, the technology developers expanded activities to create legitimacy for micro-CHP. The technology developers co-operated with each other and with potential adopters in a 'micro-CHP working group'. This working group, which was established by the intermediary organization, acts as an advocacy coalition aiming to create legitimacy for micro-CHP by lobbying for government support (function (vii)). Another example of creating legitimacy was the demonstration project that was initiated by one of the technology developers. This project did not aim at improving the technology, but at bringing micro-CHP to the attention of potential adopters and policy makers and putting the regulatory problems concerning the electricity feed-in on the political agenda (Overdiep, 2006). Thus, with respect to technology developers, perceived uncertainties stimulated the fulfillment of system functions.

In comparison to technology developers, the potential adopters turned out to be more passive. Their strategy can best be described as 'wait-and-see'. They have developed some activities in order to stay informed about the developments of micro-CHP and to represent their interest, such as participating in pilot projects and in the micro-CHP working group. However, they seem to be unwilling to make large investments in micro-CHP as long as major uncertainties remain. They delay action until the uncertainties will be reduced by others (the technology developers) or by time. Thus, for these actors, perceived uncertainties seemed to block the fulfillment of system functions.

Government. The government has been stimulating the development of micro-CHP under the framework of the 'energy transition policy'. Transition management is a new Dutch governance approach, complementary to the regular energy policy, aimed at stimulating and managing the transition towards a sustainable energy system (Rotmans *et al.*, 2001; Rotmans, 2003).[8] One of the basic assumptions of this approach, is that experiments help to deal with uncertainties about the long-term system change (Rotmans, 2003; also see the article by Voß *et al.* (2007) in this special issue). Within this policy framework, several micro-CHP experiments have been initiated (Ministerie van EZ, 2004; 2006). By stating that micro-CHP is a promising technology and by supporting experiments with micro-CHP, the transition policy of the Dutch government helps to guide the direction of the search (function (iv) in the section 'Functioning of Innovation Systems'). A strong and visible preference of the government for micro-CHP can affect the R&D priority setting positively, thus reducing uncertainties about the possibility of investing in different technological alternatives (i.e. technological uncertainty).

Guiding the direction of the search, however, is not enough to stimulate emerging transition technologies. An important task of the government is to reduce the political uncertainty perceived by the market parties. On the one hand, the government has been aware of its task to reduce uncertainties about subsidies, the electricity feed-in policy, and so on. On the other hand, however, the government has argued that market parties should realize that uncertainties due to changes in policy are inevitable and that market parties should anticipate these changes instead of calling the government unreliable. This statement points out that there is a tension between the government and the market parties with respect to who should take the lead in bringing about the uncertain transition towards sustainability.

Overall, it is concluded that the role of the government has been quite limited in this phase of the transition process. Although the government has stimulated micro-CHP in the 'energy transition policy' framework, governmental policy has not yet reduced the political uncertainties that play such an important role in the innovation decisions of the market parties. However, in this early transition phase, the expectations of micro-CHP seem high enough to counter the political uncertainties. Instead of a lack of function fulfillment, there is a significant effort by technology developers to reduce existing uncertainties by fulfilling a number of system functions. Therefore, it is concluded that, in this early phase, political uncertainty has no noticeable negative effect on the transition as a whole.

The Case of Biofuels

The second empirical case focuses on the transition towards the use of biofuels. The use of biofuels is considered a promising option for the transition towards a sustainable transport sector in the Netherlands. Biofuels are liquid fuels, produced from biomass and used for transport purposes. A distinction can be made between first- and second-generation biofuels. First-generation biofuels are produced with commercially available technologies for the conversion of sugars and canola oils into biofuels. Second-generation biofuels involve the conversion of woody biomass into biofuels; they are produced with advanced chemical or enzymatic technologies that are not yet commercially available. The advantages of the second-generation biofuels are that much higher volumes of biofuel can be obtained from one acre of land, and that the carbon emission reductions are much higher (minus 90 per cent) compared to the first-generation biofuels (e.g. minus 30 per cent) (Faaij, 2006; Suurs & Hekkert, 2005).

This case description is structured differently than the previous case on micro-CHP. Here, the focus is primarily on one type of uncertainty that has proven to be quite dominant, namely political uncertainty.[9] Several examples are discussed of (a lack of) policy efforts of the Dutch government that have led to political uncertainties for the biofuel entrepreneurs. Furthermore, the consequences of these perceived uncertainties are analyzed in terms of the activities of the various actors (especially entrepreneurs) involved in the transition. Since the perceived political uncertainty changes over time, three periods that differ in terms of political climate are analyzed. The data for this case were based on a review of grey literature (newspaper articles, professional journals and policy documents), reported in Suurs & Hekkert (2005). The literature study led to a chronological overview of activities developed by the various actors involved (i.e. governmental institutions, entrepreneurs) in the Netherlands in the period 1990–2005. From this overview, an analysis was made of how various steering initiatives of the Dutch government have influenced the perceived uncertainties and behavior of the actors involved.

Uncertainty about the General Support of Biofuels (1990–95)

The first initiatives regarding the use of biofuels in the Netherlands started in the early 1990s. A few public transport companies and local authorities initiated experiments in which they adopted the new fuels. The driving force behind these experiments was the EU's political pressure to stimulate the use of biofuels and the successful developments in Germany and France. The EU contributed to the financing of these experiments (*ANP*, 1994).

Initially, these experiments did not lead to a general take-off of the transition to biofuels. Further expansion of activities was slowed down severely by the high prices of biofuels (Rotterdams Dagblad, 2004) and by the unwillingness of the Dutch government to compensate for these higher prices (*ANP*, 1993). In the Netherlands a fierce debate took place on the desirability of biofuels. Environmental organizations and academics questioned the environmental performance of biofuels from sugar beets and canola (first-generation biofuels). On the one hand, EU guidelines forced the Dutch government to stimulate the use of biofuels but, at the same time, the government was also confronted with a lobby against the present production methods of biofuels. This created a climate in which there was a lack of clear policy regarding biofuels. Since government support of biofuels was necessary to compensate for the higher production costs, this led to a poor entrepreneurial climate. A tax reduction on biofuels would lead to competing prices with conventional fuels but, at this point in time, a general tax exemption was not political reality (*ANP*, 1993). The hope for better circumstances remained, due to increasing pressure by the EU on member states to implement policies stimulating biofuels.

Uncertainties about the future of biofuels in the Netherlands led to an agricultural lobby in support of biofuels (*Trouw*, 1995). The agricultural sector was interested in biofuel production, since farmers could collect EU subsidies for producing non-food crops, and they could generate additional turnover by selling feedstock for biofuel production. Eventually, this led to tax exemptions for some biodiesel experiments (*Het Financieele Dagblad*, 1995). These experiments were quite successful and triggered more activities in terms of lobby actions (*ANP*, 1997; *NRC Handelsblad*, 1999), research (*ANP*, 1996; *De Volkskrant*, 1998; *Trouw*, 1995) and coalition forming (functions (vii), (ii) and (iii) in the section 'Functioning of Innovation Systems'). The lobby proved to be successful when regulations for experiments with tax-free biodiesel for trucks passed parliament (*Trouw*, 2001).

Thus, the political climate in this period can be characterized by the situation that some projects received a temporary tax reduction, yet there was a general uncertainty about the potential of tax reduction for new projects and a follow-up of tax reductions when the permits granted would end. On the one hand, this political uncertainty slowed down the take-off of the use of biofuels but, on the other hand, this led to actions by entrepreneurs to influence the political climate and to experiments to show the benefit of the new technology.

Uncertainty about the Direction of the Transition Process (1995–2002)

The next transition phase was characterized by the clear preference of the Dutch government for second-generation biofuels. This preference found its origin in the strong lobby of academics and environmental NGOs against first-generation bio-fuels. With the second-generation biofuels still in the R&D stage, this period was characterized by a significant research effort, partly financed by the Dutch government (*De Volkskrant*, 1998). Thus, uncertainty about the technological feasibility of second-generation biofuels was countered by R&D activities (function (ii)) and the formation of R&D collaborations (function (iii)). The main parties involved were large vested firms with stakes in the oil, alcohol and technology development business. Only one starter was part of this process, yet this was a spin-off of the multinational Royal Dutch Shell (*De Volkskrant*, 1998). Generally, the parties involved in second-generation biofuels were not involved in first-generation bio-fuels. The only exception was Nedalco, a producer of alcohol interested in both (first- and second-generation) methods to produce bioethanol (*Duurzame Energie*, 1997; *Het Financieele Dagblad*, 1996).

For entrepreneurs involved in the first-generation technology, this new line of governmental policy created large uncertainties, since the future role of first-generation biofuels was questioned strongly and an alternative solution was offered. As a result, the progress in first-generation technology development and adoption stagnated in this period.

Uncertainty about Market Formation (2002–05)

The significant R&D initiatives led to a 'proof of principle' for several second-generation technologies. The Dutch government reserved resources for contributing to the construction of a pilot plant. However, none of the market parties showed any interest (NOVEM, 2002). The main argument of the market parties was that they perceived uncertainty about the size of the Dutch market for bio-fuels. Up to this point, the government had never put in a serious effort to create a market for biofuels by means of a general tax exemption or by setting a standard for a fixed share of biofuels in automotive fuels. In this case, the strategy of the Dutch government to invest in R&D instead of investing in a market for bio-fuels, turned out to be unsuccessful. Even though the R&D initiatives led to considerable knowledge development (function (ii)), the final step towards the market was not taken.

Meanwhile, activities related to first-generation biofuels started to pick-up again. One of the reasons for this was that the pressure of the EU on the Dutch government increased considerably. Since the technology development of second-generation biofuels proved to be too slow to meet the EU directive, the government decided to fall back on first generation and to allow first-generation technology to be part of the R&D program (Suurs & Hekkert, 2005). This proved to be a major boost in creating legitimacy for this technology (function (vii)). In addition, two players became very active in promoting first-generation technology.

The first actor was a firm called SolarOilSystems that decided to build a biofuel mill based on canola (*Bizz*, 2002; NOVEM, 2005). SolarOilSystems managed to create considerable local support from state authorities and agricultural organizations, to successfully lobby for tax exemptions and to successfully build customer networks to create total production and consumption chains. Influenced by this example and by the expectation that the EU directive would be implemented in the Netherlands in 2004, seven oil mills were built.

The other active entrepreneur was Nedalco, a Dutch producer of alcohol. Nedalco was involved in the second-generation R&D program to produce ethanol from wood, but simultaneously lobbied for better market conditions for first-generation bioethanol based on sugar beets. The firm's commercial interest in the production of bioethanol was large, due to the potential increase of its production capacity, leading to lower production costs and higher profits. Nedalco managed to lobby for a temporal tax exemption for a fixed amount of ethanol and some R&D subsidies (Suurs & Hekkert, 2005).

In 2005, the Dutch government realized it needed to put in a serious effort to come up with a long-term vision regarding biofuels in the Netherlands. Since the R&D trajectory had proved to be too slow, the government switched from R&D stimulation to market stimulation by means of a tax exemption (up to €70 million) for all biofuels in 2006 and an obligation for oil companies to mix two per cent biofuel in automotive fuels in 2007.

At this moment in time, it is too early to tell how these market stimulation instruments will influence entrepreneurs' activities. However, Nedalco has stated that this policy does not provide sufficient certainty to invest in a bioethanol plant. For Nedalco, sufficient contracts with potential customers of bioethanol will need to be signed.

This period shows a remarkable difference between the behavior of entrepreneurs under similar regimes of political uncertainty. Established firms with a small interest in these developments postpone investments, as market conditions are uncertain. Nedalco, an established producer with high stakes in these new developments, develops significant lobby activities but is unwilling to invest under uncertain market conditions. Finally, one sees small entrepreneurs, new entrants, willing to invest even though market uncertainties are considerable, and quite actively influencing their environment. One possible explanation for the difference in behavior is the required capital investment in technology. For Nedalco and the second-generation technologies, investments are exorbitant, while first-generation technology is relatively low-tech, resulting in lower investment costs.

Conclusions

The aim of this article was to come to a better understanding of the role of uncertainties in transitions, by examining two empirical cases about emerging energy technologies in the Netherlands. The following research question was posed: 'Which types of uncertainties are perceived by the actors involved in emerging technological trajectories, and how do they deal with these uncertainties?'.

The micro-CHP case demonstrates that different types of perceived uncertainties influence the innovation decisions of the actors involved. The most dominant sources of uncertainty in this case are technological and political uncertainty, followed by consumer uncertainty. Furthermore, this case shows that responses to uncertainty differ considerably between different actors.

Technology developers who have a high stake in the development of micro-CHP actively try to cope with perceived uncertainties by developing certain key activities that contribute to the functioning of the innovation system. In reaction to technological uncertainty, they initiate knowledge development activities. In reaction to political uncertainty, the technology developers initiated activities to create legitimacy for micro-CHP. In short, perceived uncertainties seem to stimulate the fulfillment of system functions by technology developers. The potential adopters of micro-CHP, on the contrary, seem to follow a wait-and-see strategy. They do develop some activities, such as participating in demonstration projects initiated by technology developers, but are unwilling to invest actively while they still perceive major uncertainties. Thus, for these actors, perceived uncertainties seem to block the fulfillment of system functions.

The role of the Dutch government has been quite limited in this early transition phase. Governmental action has not been sufficient to reduce the political uncertainties that play such a dominant role in the micro-CHP case. Despite the perceived uncertainties and the limited governmental initiatives to reduce these uncertainties, the transition process has not been hampered since technology developers have been playing a leading role and are still making progress in the development of micro-CHP. However, as discussed below, the blocking effect of political uncertainties might increase once micro-CHP becomes ready for market introduction.

Comparing the micro-CHP case to the biofuels case, one sees some remarkable similarities and differences. First—just like in the micro-CHP case—the biofuels actors react differently to perceived uncertainties. Similar perceptions of uncertainty about the size of the future market made some entrepreneurs decide not to invest in the production of biofuels, while others did invest in production facilities. The size of the initial investments and the ability of the entrepreneurs to build networks with early adoptors seem to be crucial in these decisions. It is also noticeable that the new entrants, in particular, are the ones who decide to invest, while the larger incumbent players behave in a more risk-averse fashion. This acknowledges the often-described principle that small new entrants are more capable of developing flexible strategies in fast-changing markets than large, established firms.

Another similarity is that the high level of political uncertainty in the biofuels case did not lead to a lack of key activities. In fact, many lobby activities (contributing to the system function 'creation of legitimacy') and a significant number of research activities (i.e. the 'knowledge development' function) are observed. However, compared to other countries, the number of entrepreneurial activities in the Netherlands has been quite low. Countries such as Germany, France and Austria show a much higher diffusion of biofuels than the Netherlands (Suurs & Hekkert, 2005). The large political uncertainties have blocked the diffusion of biofuels in the Netherlands, while uncertainties block crucial system functions (e.g. entrepreneurial activities). This differs from the micro-CHP case, where these patterns could not be found.

Even though entrepreneurial activities are blocked by political uncertainties, the actors in the biofuels case seem to be stimulated to develop other activities with the aim of countering these uncertainties. Political uncertainty seems to induce lobby activities in order to reduce these political uncertainties and technological uncertainties seem to lead to activities that are typical for early transition-phases, such as knowledge development. This is in line with our findings in the micro-CHP case. Micro-CHP is in an early stage of transition as well and, in this case, one also sees that activities are developed to counter specific uncertainties (e.g. knowledge development to counter technological uncertainty, and networking and lobbying to counter political uncertainty).

The biofuels case also showed that, as a technology develops further and becomes suitable for entering the market, the uncertainties seem to have a greater influence than in earlier phases of a technological trajectory. In this case, large uncertainties hamper the fulfillment of crucial system functions. A logical explanation for this phenomenon is stated by an important biofuel entrepreneur. He states that, when a biofuel firm is in the phase of entering a market with a new product, much more resources and management commitment are needed than in earlier phases. Before entering a market, a solid business-case needs to back-up investment plans and convince the management. Since large uncertainties have a major impact on the robustness of the business-case, the influence is larger in this setting than in earlier phases (Hekkert & Suurs, 2005).

It is always difficult to generalize the results of two case studies. Our findings show a number of similarities between the two cases. Since the two cases are very different in terms of involved networks and technological domain, these similarities may well hold for other emerging technological trajectories as well. The differences between the case studies may be explained by the difference in transition phase but also by case-specific circumstances. More case studies are necessary to be capable of generalizing the results to all emerging technological trajectories.[10]

What are the implications for policy? First, the empirical cases have shown that perceived uncertainties play an important role in innovation and transition processes. Secondly, in early phases of transition, the role of uncertainties seems to be less crucial than in later phases. This is an important observation for policy makers since it has also been shown that political uncertainties can hamper entrepreneurial activities greatly and thereby market introduction of sustainable technologies. Policy makers should therefore be very active in the phase just before and during market introduction in communicating well with entrepreneurs about their perception of crucial uncertainties and developing policy instruments to (temporarily) bring down the level of uncertainties. Due to the large diversity of the types of uncertainties that are being perceived and the effects of these uncertainties on the behavior of the actors involved, it is impossible to design a simple and generally applicable policy strategy. In order to deal effectively with uncertainties in transitions, it is recommended that a portfolio of various steering instruments be developed, which can be applied in different situations. Looking at the framework of steering theories proposed by Voß, Newig, Kastens, Monstadt, Nölting, 2007), this portfolio may contain elements of all clusters of steering strategies. For example, providing guidance can help to reduce uncertainty about the direction of technological development, building actor networks can reduce uncertainty about the behavior of others (such as competitors or consumers) and so on. In order to improve our ability to steer transition towards sustainability, more insight is needed into the influence of different steering strategies on actors' perceptions of and reactions to uncertainties.

Acknowledgement

This article is supported financially by NWO (Dutch Organization for Scientific Research) and KSI (the Dutch Knowledge network on System Innovations and transitions). The authors would like to thank the interviewees of the micro-CHP case. Roald Suurs is acknowledged for collecting data on the development of biofuels in the Netherlands.

Notes

1. Non-commercial adoptors are not considered entrepreneurs.
2. For example, an important consumers' characteristic for energy technologies is the energy demand.
3. The concept of 'innovation system' is a heuristic attempt developed to analyze all such societal subsystems, actors and institutions contributing, in one way or another, directly or indirectly, intentionally or not, to the emergence or production of innovation (Hekkert *et al.*, 2006).
4. The term 'key activities' is used when referring to the actor-level and the term 'functions' when referring to the system-level. If a function is being fulfilled well, this means that actors have developed many key activities that contribute to this function. To illustrate, attending a conference or organizing a workshop are examples of actors' key activities, that contribute to the function of 'knowledge diffusion though networks'.
5. All four technology developers that were, at the time, developing activities in the Netherlands were interviewed.
6. The group of potential adopters consisted of organizations that can play an important role in generating intermediary demand for micro-CHP (i.e. energy companies and housing organizations). It was not possible to interview end-users (house owners or tenants), as they have not been involved in the development activities and have not yet been made aware of micro-CHP.
7. Most of the uncertainties that the interviewees mentioned (without having knowledge of our typology of uncertainty sources in this stage of the interview) related to technology or politics. When the interviewees had to rank the uncertainty sources according to their relative importance, technological uncertainty and political uncertainty scored overall highest, followed by consumer uncertainty. Four interviewees clarified their ranking by stressing that technological uncertainty and political uncertainty were far more important in this early stage of development than the other uncertainty sources.
8. Whereas the regular energy policy is aimed at short-term goals (approx. ten years from now), the energy transition policy focuses on the long term. The energy transition policy is based on a different, more process-orientated, governance approach. Some key elements of the 'energy transition policy' involve heterogeneous actors, stimulating learning processes, and creating a wide playing field. For a comparison between the two approaches, see Rotmans (2003).
9. The alternative focus is a direct result of differences in research strategy. Contrary to the CHP case, the biofuels case was analysed according to functions of innovation systems method, as reported in Suurs & Hekkert (2005). Due to these differences in data collection methodology, we are unable to order the importance of the different uncertainties. However, the data allow us to analyse the effect of different uncertainties on entrepreneurial action. The emphasis is on political uncertainties due to a bias in data availability.
10. In the research program that led to this article, additional case studies are being performed on uncertainty perception in technological trajectories around biomass gasification and biomass combustion.

References

ANP (1993) Bukman sceptisch over kansen bio-brandstof, *ANP*, 27 May.
ANP (1994) Europese subsidie voor Rotterdamse verkeersprojecten, *ANP*, 24 February.
ANP (1996) Van Aartsen wil fiscale steun bio-ethanolproject, *ANP*, 15 March.
ANP (1997) Nedalco is wachten op accijnsvrijstelling bio-ethanol zat, *ANP*, 4 February.
Bizz (2002) De Oliemolen (2), *Bizz*, 15 November.
Corrêa, H. L. (1994) *Linking Flexibility, Uncertainty and Variability in Manufacturing Systems: Managing un-planned change in the automtative industry* (Aldershot: Avebury).
Dess, G. G. & Beard, D. W. (1984) Dimensions of Organizational Task Environments, *Administrative Science Quarterly*, 29(1), pp. 52–73.
De Volkskrant (1998) Snelle olie uit hout, *De Volkskrant*, 24 January.
Duurzame Energie (1997) Nedalco start commerciële productie bio-ethanol, *Duurzame Energie*, August.
Elzen, B., Geels, F. & Green, K. (2004) *System Innovation and the Transition to Sustainability. Theory, Evidence and Policy* (Cheltenham, UK: Edward Elgar).
Elzen, B. & Wieczorek, A. (2005) Transitions towards sustainability through system innovation, *Technological Forecasting and Social Change*, 72(6), pp. 651–661.

Faaij, A. P. C. (2006) Bio-energy in Europe: changing technology choices, *Energy Policy*, 34(3), pp. 322–342.

Foxon, T. J., Gross, R., Chase, A., Howes, J., Arnall, A. & Anderson, D. (2005) UK innovation systems for new and renewable energy technologies: drivers, barriers and systems failures, *Energy Policy*, 33(16), pp. 2123–2137.

Geels, F. W. (2002a) Technological transitions as evolutionary reconfiguration processes: a multi-level perspective and a case-study, *Research Policy*, 31(8–9), pp. 1257–1274.

Geels, F. W. (2002b) *Understanding the dynamics of technological transitions: a co-evolutionary and socio-technical analysis* (Enschede, Twente University: Centre for Studies of Science, Technology and Society).

Grubler, A. (1998) *Technology and global change* (Cambridge: Cambridge University Press).

Grubler, A., Nakicenovic, N. & Victor, D. G. (1999) Dynamics of energy technologies and global change, *Energy Policy*, 27(5), pp. 247–280.

Grunwald, A. (2007) Working Towards Sustainable Development in the Face of Uncertainty and Incomplete Knowledge, *Journal of Environmental Policy and Planning*, 9(3,4), pp. 245–262.

Hekkert, M. P., Suurs, R. A. A., Negro, S. O., Kuhlmann, S., & Smits, R. E. H. M. (2007) Functions of innovation systems: A new approach for analysing technological change, *Technological Forecasting and Social Change*, 74(4), pp. 413–432.

Het Financieele Dagblad (1995) Biodiesel tijdelijk vrij van accijns, *Het Financieele Dagblad*, 8 March.

Het Financieele Dagblad (1996) Suiker unie zet conserven in de etalage, *Het Financieele Dagblad*, 22 May.

Huang, Y. H. & Wu, J. H. (2007) Technological system and renewable energy policy: A case study of solar photovoltaic in Taiwan, *Renewable and Sustainable Energy Reviews*, 11(2), pp. 345–356.

Jacobsson, S. (2005) *Formation and growth of sectoral innovation systems—'functional analysis' as a tool for policy makers in identifying policy issues* (Gothenburg, Sweden: RIDE, IMIT and Chalmers University of Technology: Environmental Systems Analysis).

Jacobsson, S. & Bergek, A. (2004) Transforming the energy sector: the evolution of technological systems in renewable energy technology, *Industrial and Corporate Change*, 13(5), pp. 815–849.

Jacobsson, S. & Johnson, A. (2000) The diffusion of renewable energy technology: an analytical framework and key issues for research, *Energy Policy*, 28(9), pp. 625–640.

Jauch, L. R. & Kraft, K. L. (1986) Strategic Management of Uncertainty, *The Academy of Management Review*, 11(4), pp. 777–790.

Johnson, A. (2001) *Functions in Innovation System Approaches* (Aalborg, Denmark: DRUID's Nelson-Winter Conference).

Kemp, R. & Soete, L. (1992) The Greening of Technological-Progress—an Evolutionary Perspective, *Futures*, 24(5), pp. 437–457.

Koppenjan, J. F. M. & Klijn, E. H. (2004) *Managing uncertainties in networks* (London: Routledge).

Kreiser, P. & Marino, L. (2002) Analyzing the historical development of the environmental uncertainty construct, *Management Decision*, 40(9), pp. 895–905.

Lange, H. & Garrelts, H. (2007) Risk Management at the Science–Policy Interface: Two Contrasting Cases in the Field of Flood Protection in Germany, *Journal of Environmental Policy and Planning*, 9(3,4), pp. 00–01.

Meijer, I. S. M., Hekkert, M. P., Faber, J., & Smits, R. E. M. H. (2006) Perceived uncertainties regarding socio-technological transformation: towards a framework, *International Journal of Foresight and Innovation Policy*, 2(2), pp. 214–240.

Meijer, I. S. M., Hekkert, M. P., & Koppenjan, J. F. M. (2007) How perceived uncertainties influence transitions; the case of micro-CHP in the Netherlands, *Technological Forecasting and Social Change*, 74(4), pp. 519–537.

Milliken, F. (1987) Three Types of Perceived Uncertainty about the Environment: State, Effect, and Response Uncertainty, *The Academy of Management Review*, 12(1), pp. 133–143.

Ministerie van EZ (2004) Uitdagingen transitiepaden Nieuw Gas vastgesteld, *Nieuwsbrief Energietransitie*, 15.

Ministerie van EZ (2006) Website Energietransitie, thema Nieuw Gas. Available at www.energietransitie.nl (accessed 15 February 2006).

NOVEM (2002) Mogelijkheden productie Fischer Tropsch brandstof via biomassa vergassing nader onderzocht, NOVEM GAVe-mail. Vol. 9

NOVEM (2005) Nederlandse biobrandstofinitiatieven in vogelvlucht, NOVEM GAVe-mail, Vol. 4.

NRC Handelsblad (1999) 'Groene' olie kan aardolie vervangen, *NRC Handelsblad*, 7 October.

Overdiep, H. (2006) Met microWKK op weg naar een duurzamere elektriciteitsopwekking, *Bode*, 1, pp. 1–4.

Rotmans, J. (2003) *Transitiemanagement: sleutel voor een duurzame samenleving* (Assen: Koninklijke Van Gorcum).

Rotmans, J., Kemp, R. & van Asselt, M. (2001) More evolution than revolution: transition management in public policy, *Foresight: the journal of futures studies, strategic thinking and policy*, 3(1), pp. 15–32.

Rotterdams Dagblad (2004) Regio positief over schonere voertuigen; Alle gemeentelijke wagenparken over op bio-ethanol, *Rotterdams Dagblad*, 20 December.

Sabatier, P. A. (1988) An Advocacy Coalition Framework of Policy Change and the Role of Policy-Oriented Learning Therein, *Policy Sciences*, 21(2–3), pp. 129–168.

Sabatier, P. A. (1998) The advocacy coalition framework: revisions and relevance for Europe, *Journal of European Public Policy*, 5(1), pp. 98–130.

Sagar, A. D. & Holdren, J. P. (2002) Assessing the global energy innovation system: some key issues, *Energy Policy*, 30(6), pp. 465–469.

Smith, A., Stirling, A. & Berkhout, F. (2005) The governance of sustainable socio-technical transitions, *Research Policy*, 34(10), pp. 1491–1510.

Smits, R. E. H. M. & Kuhlmann, S. (2004) The rise of systemic instruments in innovation policy, *The International Journal of Foresight and Innovation Policy*, 1, pp. 4–32.

Suurs, R. A. A. & Hekkert, M. P. (2005) *Naar een methode voor het evalueren van transitietrajecten; functies van innovatiesystemen toegepast op 'biobrandstoffen in Nederland'* (Utrecht: Copernicus Institute).

Suurs, R. A. A., Hekkert, M. P., Meeus, M., & Nieuwlaar, E. (2004) An Actor Oriented Approach for Assessing Transition Trajectories Towards a Sustainable Energy System: An Explorative Case Study on the Dutch Transition to Climate-Neutral Transport Fuel Chains, *Innovation: Management, Policy & Practice*, 6(2), pp. 269–286.

Trouw (1995) Milieuvriendelijk op diesel uit koolzaad door Friesland varen, *Trouw*, 18 March.

Trouw (2001) Vracht op biodiesel, *Trouw*, 23 May.

Unruh, G. C. (2002) Escaping carbon lock-in, *Energy Policy*, 30(4), pp. 317–325.

Van Asselt, M. B. A. (2000) *Perspectives on Uncertainty and Risk* (Dordrecht: Kluwer Academic Publishers).

Voß, J.-P., Newig, J. *et al.* (2005) Steering for Sustainable Development—Contexts of Ambivalence, Uncertainty and Distributed Power. Partial draft for the workshop on Governance for Sustainable Development, Berlin.

Voß, J.-P., Newig, J., Kastens, B., Monstadt, J. & Nölting, B. (2007) Steering for Sustainable Development: a Typology of Problems and Strategies with Respect to Ambivalence, Uncertainty and Distributed Power, *Journal of Environmental Policy and Planning*, 9(3,4), pp. 193–212.

Walker, W. E., Harremoes, P., Rotmans, J., vander Sluijs, J. P. van Asselt, M. B. A., Janssen, P., & Krayer von Krauss, M. P. (2003) Defining uncertainty: A Conceptual Basis for Uncertainty Management in Model-Based Decision Support, *Integrated Assessment*, 4(1), pp. 5–17.

Weaver, P., Jansen, L., van Grootveld, G., van Spiegel, E. & Vergragt, P. (2000) *Sustainable Technology Development* (Sheffield: Greenleaf Publishing).

Wernerfelt, B. & Karnani, A. (1987) Competitive Strategy Under Uncertainty, *Strategic Management Journal*, 8(2), pp. 187–194.

Who is in Charge here? Governance for Sustainable Development in a Complex World

JAMES MEADOWCROFT

This paper explores one of the major challenges associated with governance for sustainable development: managing change in a context where power is distributed across diverse societal subsystems and among many societal actors. The discussion is divided into four parts. The first examines the idea of 'governance for sustainable development'. The second considers the diffusion of power in modern societies. The third explores the extent to which this constitutes a problem for sustainable development. The final section advances some approaches to governing for sustainable development in a radically 'decentred' societal context.

1. Governance for sustainable development

In the most straightforward sense, 'governance for sustainable development' refers to processes of socio-political governance oriented towards the attainment of sustainable development. It encompasses public debate, political decision-making, policy formation and implementation, and complex interactions among public authorities, private business and civil society – *in so far as these relate to steering societal development along more sustainable lines* (Meadowcroft, 2005).

Since it was brought to international attention by the Report of the World Commission on Environment and Development in 1987, 'sustainable development' has been linked with a series of normative ideas including: protection of the environment, particularly the essential life support functions of the global ecosphere; promotion of human welfare, especially the urgent development needs of the poor; concern for the well being of future generations; and public participation in environment and development decision making. It is often spoken about in terms of achieving an appropriate balance between three 'pillars' – the environment, economy, and society. Basically, it is about reorienting the development trajectory so that genuine societal advance can be sustained.

Sustainable development is a complex and contested concept, and despite the pages of 'consensus documents' adopted by international agencies and conferences, there remain many different perspectives on what it entails and the scale of reforms required to give it force. In academic circles there have been recurrent debates about whether it represents a rigorous philosophical or economic concept, and the difficulty of translating it into specific policy prescriptions. Although such debates have generated insights, they have often missed the critical *political* point that this concept was *not* formulated as part of the technical vocabulary of social science, or as an operational rule that would allow policy outputs to be automatically read off from a list of situational inputs. Rather it was designed as a normative point of reference for environment and development policy making (Lafferty, 1996; Jacobs, 1999). Like other political concepts – such as 'liberty', 'democracy' and 'justice' – it helps to frame and focus debate, while being open to constant interrogation and re-interpretation.

Turning now to 'governance', it is worth noting that the term is not new to the political lexicon. But over the past twenty years it has increasingly found favour in discussion of the changing responsibilities of public authorities, and the varied ways order is generated in modern society. Contemporary usage has been shaped by debates about 'good governance' that emerged in international development circles in the late 1980s (World Bank, 1991; DAC-OECD, 1993). Although the formulations used by specific international bodies differ, 'governance' is today generally understood to refer to practices through which societies are governed, and 'good governance' is associated with a diverse array of criteria including: effectiveness and efficiency, the rule of law, participation, accountability, transparency, respect for human rights, the absence of corruption, toleration of difference, and gender equity (UNDP, 1997; Plumptre and Graham, 1999).

Within political science interest in 'governance' has been linked with attempts to understand changing patterns of state/societal interaction. Some theorists have associated 'governance' with *new* forms of socio/political interaction. Thus Rhodes defined 'governance' in terms of 'self-organising inter-organisational networks' that constitute 'an alternative to, not a hybrid of, markets and hierarchies' (Rhodes, 1996, p. 659). In a similar vein, Jessop has spoken of 'governance' as a form of social co-ordination based on 'dialogic rationality', where goals are 'modified in and through ongoing negotiation and reflection' (Jessop, 2000, p. 17).

In contrast, others have invoked governance as a more general term that embraces *the different ways* in which co-ordination is achieved. Kooiman describes 'social-political governance' as 'arrangements in which public as well as private actors aim at solving societal problems' (Kooiman, 2000). 'Diversity', 'complexity', and 'dynamics' are understood as the outstanding features of modern society, and they impel the development of more varied governance practices involving hybrids of three basic governing forms – 'self-governance', 'co-governance' and 'hierarchical governance' (Kooiman, 2003). A similarly open textured notion of 'governance' is employed by Pierre and Peters, who relate it both to institutional structures and interactive processes (Pierre and Peters, 2000).

How do such broad debates about 'governance' relate to sustainable development? With respect to the notion of 'good governance', one might say that concern with sustainable development is *one element* of what it means to practice 'good governance'. Sustainable development is an internationally recognised goal which governments (and other organisations with governance responsibilities) *ought* to pursue. On the other hand, one could argue that 'good governance' is necessary *in order to achieve* sustainable development. It was in this vein that the Introduction to the 'Plan of Implementation' adopted at the 2002 World Summit on Sustainable Development declared: 'Good governance within each country and at the international level is essential for sustainable development' (WSSD, 2002, p. 8).

Yet neither formulation takes one very far. The first is simply an assertion that 'good governance' embraces all internationally legitimated norms, while the second presumably holds true for *any* substantive societal end – for without 'good governance' how can we be sure to attain desirable social goals? Of course, it is true that to the extent that debates about 'good governance' have established the sorts of governing practices that encourage sound development, then these are also relevant for sustainable development. But such a general linkage fails to direct attention to the *specificity* of the challenge of sustainable development.

With respect to debates in political science it is the wider, rather than the narrower, usage of 'governance' cited above that can most fruitfully be linked with sustainable development: for our interest is in the entirety of governance processes and mechanisms as they may be adapted to promote this social aim. And many insights from existing governance literatures are relevant to 'governance for sustainable development': for example, discussions of emerging mechanisms of global rule, shifts in the structure and functions the state, and inter-actions among forms of 'self', 'co' and 'hierarchical' governance.

All this may have left the reader with the impression that 'governance for sustainable development' is impossibly broad. After all, we are dealing with an overarching social objective that embraces economic, social and environmental dimensions. Employment policy, fiscal management, the health care system, pension arrangements, housing policy, immigration law, the fight against crime, the management of natural resources – all have some bearing on sustainable development. And most of these are reflected in the indicator sets governments are elaborating to track sustainable development (DEFRA, 2005). Does it follow that governance in every social domain should now be understood as subsidiary to some all-encompassing process of 'governance for sustainable development'? Certainly it can be viewed this way. But a more promising approach emphasises the specific 'value added' of the concept of sustainable development – *the distinctive light* which this perspective can shed on various areas of societal governance, and *the particular adjustments* that are necessary to orient social development along more sustainable lines. On this reading, 'governance for sustainable development' is primarily associated with issues central to the sustainability problematic. Three tasks stand out: (1) the identification and management of critical threats to sustainability; (2) the integration of sustainability into general practices of governance; and (3) the organization of collective reflection and decision with respect to reconciling social priorities and orienting the overall development trajectory. Here *the emphasis is not so much on the totality of socio/political governance, as on the reform of that totality in light of sustainable development.*

A number of recent studies have begun to explore explicitly the issue of governance for sustainable development (OECD, 2002a; OECD, 2002b; Lafferty, 2005; Kemp, Parto and Gibson, 2005), and it is worth noting two important points that can be distilled from this work: one relates to the scale of the necessary social transformation, and the other to the character of the implied steering logic.

First, there is little doubt that *sustainable development represents an ambitious agenda for societal change*. To reconcile continued economic and social improvement with the preservation of global ecological systems, will require a dramatic 'decoupling' of economic activity from environmental loading (OECD, 2001). In global terms the significance of this 'decoupling' is particularly evident with respect to climate change, where recent scientific assessments suggest that stabilisation of the climate system will eventually require a decline in global carbon dioxide emissions to *a small fraction* of current levels (IPCC, 2001; Schellnhuber, 2006). But threats to long term ecological integrity are manifest in many other areas including water use, the management of forests and fisheries, patterns of land utilisation, chemical releases, and the disposal of wastes (EEA, 1999; MEA, 2006).

Addressing this challenge will necessitate a radical shift in existing patterns of production and consumption, and the transformation of major socio-economic sectors including energy, transport, agriculture, manufacturing and construction (Toner, 2006). Some commentators have dubbed this 'the next industrial revolution' (McDonough and Braungart, 1998; Hawken, Lovins and Lovins, 1999). The analogy is apt since it suggests not only fundamental technological advance, but also profound shifts in the organization of society. In this respect engagement with the politically charged issue of consumption cannot be deferred indefinitely (Princen, Maniates and Conca, 2002).

Second, the idea of *governance for sustainable development embodies a specific 'steering logic'*. Sustainable development is not a spontaneous social product: it requires goal-directed intervention by governments and other actors. The objective is to displace the direction of social movement (to a certain extent, and in certain dimensions) so that current (authentic) developmental priorities are attained, while the preconditions for subsequent social advance are not eroded (Meadowcroft, 1997). This form of 'steering' does not seek to control every dimension of social life. It can accept that the future is largely unknown and unknowable, and recognise that our collective capacities to determine what is to come are limited (Meadowcroft, 1999). Each generation makes its own choices; and since generations overlap, orientations towards the future are constantly reviewed. But even in the face of this radical uncertainty and indeterminacy human beings can try: 1) to orient society towards the attainment of desirable objectives and the avoidance of dangerous pitfalls; 2) to take action to protect groups that are especially vulnerable to the unfolding pattern of change; and 3) to re-order social institutions so that they are better placed to cope with whatever the future does bring.

But who is to do this 'steering'? In a fundamental sense, governance for sustainable development implies a process of *'societal self-steering'*: society as a whole is to be involved in the critical interrogation of existing practices, and to take up the conscious effort to bring about change. Thus it involves not only actions and policies to orient development along certain lines, but also the collective discussion and decision required to define those lines. Value choices – about the kind of society in which we want to live, about the kind of world we want to leave to posterity – lie at the heart of governance for sustainable development. At base, it is not a technical project, although technical expertise is essential, but a *political* project. For, while the concept indicates issues that should be of concern, *its practical bearing cannot be established independent of the concrete life circumstances of a particular society* and the needs, interests, values and aspirations of its members. Thus governance for sustainable development is 'interactive', not just in the instrumental sense that societal inputs can facilitate progress towards known objectives, but also in the deeper sense that the objectives themselves must be collectively defined, refined and re-defined.

We shall return to consider the implications of the *reflective* and *interactive* dimensions of governance for sustainable development in the final section of this essay. Here it is important to emphasise that notwithstanding the 'society-embracing' character discussed above, this steering logic also implies an important role for public authorities at all levels – including local and regional governments, national states, supranational unions, and international bodies. In other words, 'government' is central to 'governance' for sustainable development.

In this context the continued significance of country states requires explicit acknowledgement. Despite changing patterns of societal interaction, increased international economic interdependence, and some surrender of sovereignty to supranational institutions, states remain potent actors (Pierre and Peters, 2000; Eckersley, 2004). Above all, their mechanisms of representation and democracy allow states to claim to act legitimately for the common good. Sustainable development involves the deliberate adjustment of the conditions of social life; and if it requires an interactive and participative element, it also requires formal structures to consider options and to provide for closure and authoritative decision. Thus states provide *both a means to shape 'society', and a framework through which 'society' can influence the orientation of that shaping.*

2. The problem of distributed power

That power to influence outcomes is widely distributed across society is an axiom of contemporary political life. In representative democratic political systems, with privately owned productive assets and market-mediated economies, no single group holds a monopoly on power. Much of the literature of political science is concerned with how economic and political powers interact, and with the ways different actors mobilize resources to secure their ends. Classical pluralist theories emphasised diverse forms of power and the capacity of groups to organize to articulate their interests; Marxist and neo-Marxist analyses typically stressed class conflict and underlying economic interests; while more recent writings (for example, by post-modern thinkers) have dwelt on the fractured characteristics of contemporary political life.

Although the boundary between the economic and political spheres in modern democratic societies is permeable, and has varied over time (for example, with nationalization and de-nationalization), at any given moment it poses important constraints on what actors can achieve. Of course, the diffusion of power does not stop here. Political authorities are themselves fragmented. There is vertical 'layering' of political jurisdictions extending from local governments, through regional (state, provincial or county) administrations, up to national states, and institutions of international governance. Nation states remain the critical node in this system, but sub-national governments often wield substantial powers, as do certain international bodies. The European Union represents an emergent transnational governmental form. And there are also hybrid centres of authority that deal with issues that cut across jurisdictional boundaries (for example, water resources in the Great Lakes, or tourism in the Alps). Tensions and struggles among all these myriad 'layers' of authority are a mainstay of modern politics.

'Horizontal' divisions within governments have also become more complex: various ministries and departments, specialized in particular functions, reflect the diversity of tasks government undertakes. Recent decades have also witnessed reforms to the public service that have created an array of semi-autonomous agencies with regulatory, administrative, and service delivery functions. And, as governments have increasingly turned to more 'interactive' modes of governance, involving collaboration with other societal partners (businesses, scientific and educational institutions, non governmental organizations), still more players have joined the scene. Of course, political power has long been split among constitutionally mandated authorities (executive, legislature and judiciary), as well as among political leaders, party activists, and the electorate – who are each more or less potent in particular circumstances. The media, too, exerts significant influence, particularly on how issues are framed for public debate.

The economic realm is no more straightforward. While in some contexts the proportion of economic activity controlled by the largest firms has increased, the differentiation of activity has also grown. More of the world has been drawn into the international economic system, and a greater number of internationally based firms are active within national jurisdictions. Although the business world some-times speaks with one voice, it is also riven with contradictions: the interests of various sectors, industries, and firms pull in different directions. So again, centres of power are multiple; and to some extent fluid. And of course, there are many other social groups, movements and organizations – from trade unions to religious organizations, from grass-roots campaigns to learned societies, which have specific power resources that can be applied in certain contexts to influence the course of events.

There are other ways to conceptualize the diffusion of power in the modern world. Writers in the tradition of critical theory have long noted the different 'logics' appropriate to different societal subsystems – law, science, politics, the economy, and so on (Habermas, 1996; Dryzek, 1987). Mastering these rationalities can be understood as a pre-requisite to exercising power in different domains. Taking this further, Niklaus Luhmann has insisted on the relative insularity of societal subsystems, which resist inputs from adjacent systems operating with different logics (Luhmann, 1989). The suggestion is that general social steering becomes problematic as functionally adapted subsystems operate on their own internal principles, perhaps to the detriment of the social whole.

This points to the enduring issue of 'structure' and 'agency'. In the modern polity there is both a greater number and diversity of agents that are able to influence events (either consciously, or indirectly as they pursue their own ends), and also a more complex intermeshing of 'structures' that enable and constrain these agents, but which are produced and reproduced through their actions. When considering power it is also important to keep in mind that there is a distinction between power to influence outcomes within a given system of interactions, and the power to influence the transformation of that structure into something different. There is also a difference between the capacity to block such a structural transformation (veto), and the power to affect a lasting move towards a new and relatively stable outcome.

Returning to sustainable development, it is fairly clear that what is of particular interest is the power to change social practices in the direction of sustainability (as opposed, for example, to the power to secure a particular distribution of social goods, to perpetuate a particular social or political state, or to realize a particular societal blueprint). But this means dealing with distributed centres of power. After all, problems of sustainability typically *cut across* functional administrative divisions, established territorial jurisdictions, and the economy/politics divide. They involve many actors and have implications for various societal subsystems.

We will return shortly to strategies for negotiating such a complex landscape. But first it is necessary to set this discussion of the diffusion of power in a wider context.

3. Contexualizing the problem of distributed power

The problem of distributed power is just one of three critical challenges facing efforts at governance for sustainable development identified by the editors of this special issue. Indeed, they present this as something of a conundrum. For they point out that sustainable development implies some form of 'steering' – to ensure that societal development avoids 'unsustainable' outcomes. To steer successfully we have always been told that one needs a) clear goals, b) a good understanding of relevant causal relationships, and c) the power to influence outcomes. Yet in the context of sustainable development each of these three requirements appears problematic. Goals are vague, ambivalent, or conflicted. We are plagued by uncertainties and ignorance, and power is distributed among many actors and across many subsystems. In other words, we are not sure, or we cannot agree, exactly where we want to go. We do not fully understand complex, evolving, and interlinked natural and social systems, and power is so broadly dispersed that policy-makers lack the ability to make things happen.

Of course, one could argue that this overstates the case. True, sustainable development is a complex normative standard, and yet the concept is not so abstract as to be devoid of meaning. While we can not agree on everything, enough of us can agree on what needs to be achieved to move forward (see Walker this issue). Uncertainty and ignorance beset decision-making for sustainable development. But this implies that it is particularly important to elaborate strategies a) to acquire more knowledge in critical areas, and b) to take wise decisions in the absence of full knowledge (see Grunwald this issue). Finally, while there is no doubt that power is distributed widely in modern societies, states retain considerable clout. Moreover, new linkages among governance structures and new modes of public/private interaction are evolving to cope with changing circumstances. Governments still succeed in getting things done, and they can influence the orientation of societal development

Moreover, it is important to note that these three problems express dilemmas that have *always* confronted political leaders, particularly in democratic polities: governments are bound to face multiple, conflicting, and perhaps incommensurate objectives; there will never be enough causal knowledge to tackle the issues that really worry us; and policy-makers must by definition deal with alternative centres of power. Of course, we may argue that the situation facing modern leaders is more acute than that which confronted earlier generations of decision makers. In more complex and globally interconnected societies goals are multiple and conflicted, pathways of natural and social causation are more opaque, and power is more widely diffused. On the other hand, we enjoy many advantages over earlier societies. We understand more about natural and social processes; our technologies are more powerful; and we can reflect upon past experience and our own conduct in ways that were not available to earlier generations. Thus it is not evident that we are worse placed to shape our own destiny. Like earlier peoples we make our history in conditions that are not of our choosing (Marx, 1852). Thus, to some extent, the three challenges formulated above restate general problems of democratic governance – but they do so in a context where global ecological problems and a consciousness of global inequalities have come to the fore.

This is not to suggest that these are not real challenges for governance for sustainable development. To attempt to shift the societal development trajectory, and to radically reduce the human impact on the global ecosphere, is an ambitious goal. It will involve deployment of particular stratagems to develop operational objectives, to take decisions in conditions of uncertainty, and to drive forward reforms in circumstances of distributed control.

Focusing more particularly on the issue of distributed power, several additional points need to be made. First, before we go too far down the road of bemoaning this dispersal of power, and the plethora of societal actors with which modern governments must engage, we should pause to consider the alternatives. Instead of being distributed, power might be concentrated – either exclusively in the hands of political authorities or perhaps those of private interests. Would this make the situation any better? The experience of the twentieth century suggests not. Whenever political and economic powers have been fused, or power has been monopolized by concentrated political or economic elites, all sorts of abuses proliferated. Certainly the environment did not fare well. So probably we should be reasonably content that power in modern societies is somewhat diffuse.

This observation could also be expressed in terms of the three difficulties discussed above. For if governance for sustainable development really takes place in a context where a) goals are unclear and b) knowledge is limited, should we not be glad that c) power is also dispersed? Again the alternative is worrisome. In fact, this dispersal provides some guarantee that in pursuit of 'sustainable development' governments with goals that are confused (or that represent only sectional interests), and which are armed with inadequate and erroneous understanding of natural and social causation, do not have the undisputed power to inflict their projects on the rest of society.

Indeed, just raising the idea of doing without the plural centres of power with which we are so familiar reveals how impossible this would be. To strip out the functional differentiation between economic, legal, scientific, and political spheres and discourses; to move back from specialization and differentiation of labour and knowledge; to do away with fragmented jurisdictions, the complex interpenetration of state/society, and the multitude of self and co-regulating processes; would be to attempt to undo what makes modern democratic societies distinctive. The forces leading to this diffusion of power (as well as some leading to new concentrations of power) – including functional specialization, democratization, more widespread economic development, and globalization – are among the most pervasive currents of the modern world.

So diffuse power is something we have to live with, and more than that, it is something closely associated with the *positive* features of development. Indeed, rather than thinking about the diffusion of power as a problem for sustainable development, we can also see it as an opportunity. It establishes many channels for information to flow, and many ways for feedback from social/environmental interactions to be articulated. It opens the door for multiple routes of intervention in order to encourage the turn towards sustainable development. It provides a range of different groups – all with some power resources – from which to build coalitions for change. Of course, in any given context the proliferation of centres of power, the growth in the number of implicated actors, or an increase in veto points, can make organized efforts for reform more difficult. The division of authority in federal systems is a case in point. There is clear evidence that federal systems have more difficulty coordinating country wide engagement with sustainability (Lafferty and Meadowcroft, 2000). On the other hand, federalism evolved as a political response to specific issues (self rule in contexts of large territories, and/or strong regional identities, and/or cultural differences), and without it federal states might long since have broken up – and this would now pose co-ordination problems of another sort. Moreover, federalism allows experimentation and provides a context where sub-national units can act to address issues that are not yet 'mature' on the national scene. And here decentralism has favourable impacts. Thus 'diffuse power' brings advantages as well as disadvantages. To some extent it only becomes an over-riding concern if we assume that governments (or perhaps some other social group, such as scientists or environmentalists, who would like to get governments to do their bidding) *already* know *exactly* what needs to be done and how we should go about doing it. But no one possesses that knowledge in any detail. Nor is it possible to possess this knowledge in advance of the complex societal interactions involved in collectively exploring alternatives, making interim choices, experimenting with reforms, drawing lessons, readjusting goals, and so on, that sustainable development implies.

This points to the mistake of understanding the 'steering' involved in 'governance for sustainable development' by making too direct an analogy with a classic policy 'implementation' logic. Sure, if one is concerned with addressing a particular issue (say at the program level), officials need clear goals, an adequate causal theory, and substantial implementation potential. For the presumption is that the orientation of policy has already been decisively fixed by political principals operating in a democratic context. But sustainable development is not a 'particular issue' of this kind. Rather it is a normative standard that serves as a meta-objective for policy. It must be given substance in *every* specific context, and in the process of rendering its bearing 'concrete', discrete goals, relevant causal understandings and enabling strategies can be articulated. Moreover, as was argued earlier, governance for sustainable development implies a process of *societal* self-steering. While it is important that governments can get things done, what ultimately matters is that the result of the interactions among the full array of societal actors impels development along a sustainable trajectory. In other words, the 'steering' in governance for sustainable development can be the result of actions and interactions among many actors – including those outside government, and the successful deployment of such 'distributed' or 'de-centred' steering seems fundamental to progress towards sustainable development.

To put this another way – a critical component of the 'steering' involved in governance for sustainable development are the societal interactions which can help define 'clear goals' and develop better causal theories. In this context the diffusion of power is important – because it is essential to the operation of processes whereby goals can be formulated (democratic interaction) and knowledge developed (social and scientific discourse).

Third, while the diffusion of power (among a greater or lesser number of actors or subsystems) is significant, attention must also be paid to the pattern of its distribution (how it is spread across the relevant actors and subsystems), as well as to the character of that power and the resources upon which it is based. Power is not distributed equally; instead it is lumpy. Some groups and individuals possess a lot of it, and others less. Power comes in many forms, and some of these are not readily fungible (convertible into other forms of power). Money can do a lot, but it can not do everything. The same goes for influence, knowledge, laws, arms, and all the other resources that underpin power. Generally speaking, groups with the most power are also those that have gained (or at least believe that they gain) the most from existing ways of doing things. So they are understandably resistant to certain types of change. In business, incumbent groups controlling established technologies enjoy structural advantages over later entrants and emergent technological options. Established producer groups may have secured extensive subsidies and political protections (for example, agriculture in many developed states, and much of the fossil fuel industry), and the concentrated clout of such special interests systematically overrules the diffuse interest of the general public. Alternatively, some disadvantaged groups may possess little power under present socio/political circumstances, but have the potential to exert pressure should they be mobilized.

In fact, diffusion and distribution are closely inter-related; as one shifts so does the other. But the implications of diffusion may be quite different depending on the character of the resultant distribution. A shift from a situation where power is split between two main stakeholders to one where three, four or more groups are involved (i.e. 'diffusion') *may actually make initiation of change easier* (rather than harder), depending on the particular distribution and forms of power available to each actor.

5. Steering in contexts of distributed power.

Although governance cannot be reduced to the actions of government, and sustainable development involves processes of 'societal self-steering', much of the responsibility for actually ensuring that progress is being made rests with governments. After all, governments are the only institutions with a general mandate to promote the public good with (at least in democratic systems) clear lines of accountability to the general population. Thus governments must come to terms with steering for sustainable development in the radically polycentric environment described above.

Navigating in such complex contexts – advancing an ambitious reform agenda when power is fragmented among many actors and subsystems – implies a turn towards a more *interactive/reflective* mode of governance. What does this mean? It means extending an approach to governance that consciously employs *interactions* with other power centres to define and realize goals, and that encourages *reflection* (within government but also across society) about societal circumstances in order to reassess practices and adjust initiatives. Critical elements of such an approach include:

- acting from the understanding that government is just one (albeit one crucial) component of the overall process of societal governance. Thus government actions are oriented to increase the likelihood that *the system as a whole* will evolve in the desired direction. The focus is on *enabling* those factors and forces – within the political/legal/administrative sphere *and* in society more broadly – that tend to promote a sustainable orientation towards development.

- exploiting interactions among actors to gain knowledge about interests, perspectives, and capacities and to learn about the character of societal/environmental linkages, as well as the opportunities for (and obstacles to) change. In many cases it is only by initiating action, gauging the reactions of other parties, and assessing societal impacts, that it is possible to gain the understanding necessary to define a pathway to more significant reform.

- establishing long term objectives that operationalize sustainable development in the specific societal context. Formulating such goals, as well as possible routes to their attainment and shorter term objectives to get the process of change under way, are critical to providing actors with a vision of how the system can be expected to evolve, as well as to establish reference points with respect to which progress can be assessed, and subsequent adjustments to the goals and the means applied to secure them can be oriented.

- supporting the extension of co-governance networks around specific issues, particularly those that draw together organizations from across the state/ business/civil society divide. In some cases governments may participate directly in such co-operative management regimes. But whether or not they are involved directly, governments can promote political conditions favourable to such initiatives and have a responsibility to monitor their results.

- ensuring the development of varied institutions to track social and environmental trends, to analyse and assess existing practices and the effectiveness of policy initiatives, and to audit performance. Such knowledge is important for government, but also for other organizations and for citizens. Much of this work needs to be done by a diversity of bodies supported by public resources but enjoying substantial independence from political and administrative interference.

- promoting the emergence of a vibrant 'public sphere', to accommodate continuous discussion of social choices and critical reflection on the development path and policy approaches (Torgerson, 1999). Such a dynamic space can encourage the consolidation of new ideas and practices, strengthen a public ethos of sustainability, and renew political support for the continuing process of reform.

- encouraging the growth of 'ecological citizenship' – among individual and collective actors. For, if the state can furnish both a framework for collective decision-making about sustainable development and a mechanism to actualize these decisions, it is society that will provide the underlying dynamism, inventiveness and normative impetus. This is as true for choices made in the spheres of domestic consumption and social production as it is in those of community service and political activism. So, by promoting forms of citizenship that think critically about social/environmental interactions, engage practically with collective problems, and assume responsibility for conduct in private and public life, government can strengthen the societal foundation for the transition towards sustainable development.

Interactive/reflective governance does *not* mean that government simply responds to whichever social group 'shouts the loudest' in a given context. On the contrary, interactions are used as a deliberate tool to develop knowledge and to construct an understanding of the public good that transcends particular interests, and can be shared widely. In other words, interactive/reflective governance *is a dynamic posture, oriented to exploiting the diffusion of power to promote adjustment of the development trajectory.* It is about judicious interventions to channel social energies down pathways conductive to sustainability.

Moreover, the fact that 'co-governance' arrangements are particularly appropriate does not mean that all governance is to be remade upon this model. On the contrary, more traditional modes of state action – where government proscribes and enforces – must underpin the elaboration of collaborative initiatives (Glasbergen, 1998), and 'self-governance' forms, where sets of social actors take responsibility for ordering their own affairs and maintaining standards that promote a broader social good, are also important. But such approaches can be closely inter-linked with collaborative mechanisms.

Interactive/reflexive governance, and the increasingly complex patterns of organizational interdependence which it both reflects and encourages, does not imply that governments decline to act unilaterally. On the contrary, unilateral action – passing a law, raising a tax, introducing an institutional reform – is often necessary to break the resistance of entrenched interests and to upset the existing equilibrium. Deliberate destabilization of established ways of doing things – even in the absence of the possibility of imposing a new equilibrium – can be a crucial resource of reformers. In political terms, the trick is to initiate change that creates pressures for further reform and adjustment, rather than playing into feedback mechanisms that revert to the status quo.

In recent years the idea of 'reflexivity' has been invoked by social theorists in many different ways. Bhaskar identified it with the requirement for every modern theory to locate itself within the context which brought it forth (Bhaskar, 1993). Giddens talked of 'the use of information about the conditions of an activity as a means of regularly reordering and redefining what that activity is' (Giddens, 1990, p. 86). Beck associated 'reflexive modernization' with processes 'in which one kind of modernization undercuts and changes another' (Beck, 1994, p. 2). More recently, and in the context of the discussion of governance for sustainable development, Voß and Kemp have attempted to distinguish 'first' and 'second order' reflexivity, with the first referring to the feedback effects which originally interested Beck (as modernity confronts side-effects of its own advance), and the second denoting conscious reflection that transcends the routine problem solving and fragmented rationalities of modernity and attempts to come to grips with underlying problems (Voß and Kemp, 2005). 'Learning' and 'immanent critique' are the elements highlighted by Eckersley in her book on the 'green state' which points to varying degrees of reflexivity embedded in different forms of re-appraisal and adjustment of the state's role (Eckersley, 2004, p. 81).

The understanding of interactive/reflective governance employed here actually falls closest to the orientation of Grin and Hendriks (this issue), where reflexivity is linked to the 'transformation of the governance system itself' and the search for innovative solutions to social problems by moving beyond surface manifestations to uncover structural and systemic underpinnings. There is a link to notions of 'reflexive designing' (Grin et al, 2004), and the emphasis is placed more centrally on politics, civil society, and the interstices where 'different discursive spheres overlap'. (Grin and Hendriks, this issue; see also Meadowcroft, 2004). Nevertheless, the double-barrelled term 'interactive/reflective' has been employed here to highlight two issues. The first is that active and agent-centred processes of *reflection* are the primary concern. Governance for sustainable development implies conscious reflection about the past, present and future, including the potentially radical interrogation of the nature of social progress and the character of authentic development. The second is that particular kinds of *interactions* are central to steering society along sustainable lines. On one level, this is due to the diffusion on power and the increased societal complexity considered above. But on another it is related to the democratic content of sustainable development. For without democratic and participatory interactions there is no way to generate the 'reflectivity' necessary to re-orienting the development trajectory along sustainable lines.

In attempting to influence reform initiatives across a range of social actors and sectors, interactive/reflective governance can employ many techniques. One critical element is what might be described as *'ideological steering'*: the development of sets of inter-related ideas that influence activities at all levels. Here co-ordination is achieved not by hierarchical control, but because actors (including both individuals and institutions) internalize certain perspectives, and independently orient (and re-orient) their actions in consequence. 'Ideology' is not used in the sense of classic 'political ideologies' – liberalism, conservatism, socialism, and so on. But rather to refer to broader sets of social ideas that to some extent transcend these political currents, and provide a frame within which political argument takes place. Sustainable development is clearly an idea of this type. But a concept at such a high level of abstraction needs to be surrounded by many other concepts that concretize its orientation and articulate a more complete societal vision. Codes of behaviour, standards of appropriate conduct – that apply to public officials, but also individual citizens – provide an important mechanism for influencing outcomes. Such 'universalizing' ideas can to some extent cut across the divisions among subsystems, affirming general values and orientations, and encouraging their internalization across the range of diverse societal activities. Thus education and communication strategies become critical elements in governance for sustainable development, and one should consider the role which all sorts of groups (including professional and occupational bodies) can play in developing and transmitting such 'ideology'.

In developing an interactive/reflexive approach governments can intervene to shift the distribution of power in ways that encourage adjustments conducive to sustainable development. Techniques for shifting power balances include:

- *Adjusting legal rights and responsibilities.* Government intervenes to adjust the legal obligations of established actors, making some avenues of development easier and other more difficult (costly, contentious, and risky) to pursue. An example is provided by disclosure requirements that oblige industry to identify hazardous substances used at local facilities. By requiring companies to make such information public, the balance is tipped slightly towards groups campaigning against toxic releases. Changes to liability regimes for environmental damage can work in a similar way. Or one could consider re-defining the responsibilities of regulatory agencies.
- *Creating new institutional actors.* Governments lend support (financial, organisational, moral) to encourage the creation of autonomous actors who can promote change. This can include transferring functions from the core of government to bodies that work at arms length. This can provide many advantages including reducing day to day political interference, freeing the bodies from bureaucratic routines and mind-set, increasing public confidence, and serving as a hedge against changed political priorities. New actors can be created by

hiving-off existing structures, fusing discrete units, or starting from scratch. The resultant body can be granted different degrees of autonomy. Alternatively, governments can assist other parties to organize themselves – to promote particular causes or regulate their own affairs. The encouragement of sectoral self-organization to engage with climate change is a case in point.

- *Establishing new centres of economic power.* Government intervenes to strengthen economic actors whose activities point in the direction of desired social ends. The best example here is encouragement of the green business sector (renewable energy, organic farming). Not only do such interventions secure direct environmental and economic gains (more green energy, growing export markets), they have the indirect effect of bolstering constituencies advocating further change (for example, the removal of subsidies to fossil fuels, reduction of pesticide usage). Current economic interests (with established facilities, technical know-how, jobs, export earnings, and so on) are like large masses in a gravitational system – they exert political 'pull' in proportion to their size. To counter-balance their influence it is necessary to build rival enterprises that can offer jobs, tax revenues, exports, along a line of advance congruent with sustainable development.

- *Encouraging inter-organizational collaboration.* Government encourages new patterns of interaction among existing organizations to favour innovation for sustainable development. By bringing groups together to address particular problems, power relations can be subtlety adjusted (see Kemp, this issue). These may be groups enmeshed in an established problem matrix; but by encouraging interaction in a collaborative, solution-oriented, framework, it may be possible to redefine issues and interests. Or groups previously unaware of each other's existence may be brought together in 'innovation networks' that link some combination of consumers, producers, administrators, entrepreneurs, and researchers to accelerate technological and social innovation. Such organisational alliances can encourage the development of new products and markets, new ideas and cultural values – and strengthen the resources available to those championing sustainable development.

6. Conclusion

Steering in a polycentric environment presents a real challenge for modern governments; one that is particularly evident in relation to sustainable development. Yet governments are not bereft of approaches that can fruitfully be applied. Central to this effort is the turn towards more interactive/reflective modes of governance that encourage societal 'self-analysis' to assess and reassess the development trajectory, and exploit interactive processes to increase the manageability of long term social change. Such approaches appreciate the reality of 'distributed control' and exploit it to define, and to achieve, social ends. Interestingly, key elements of this perspective on governance can also contribute to handling the two other vexatious problems discussed in this special issue: goal ambiguity and uncertainty. The interactional approach to accumulating social knowledge, the extension of co-governance arrangements, and the establishment of a vibrant public sphere, can help build consensus around societal goals – so reducing goal-centred ambivalence. The establishment of longer term operational goals, the extension of co-management arrangements, and the development of pluralistic networks for reviewing social and environmental trends and policy performance, can help reduce uncertainties.

One final point should be stressed before this discussion is brought to a close. And that is the irreducibly *political* (and ultimately democratic and participatory) character of sustainable development. Reference has already been made to the fact that governance for sustainable development is not just a technical/administrative challenge. It is not something that competent officials can get done quietly and efficiently, out of the public view, as citizens go about their everyday business. On the contrary, it is an inherently political process – for it requires societal decisions about desirable ways of life, and about how benefits and burdens are to be shared among different communities and different generations, and between humankind and other inhabitants of this planet. The sort of radical decoupling of economic activity from environmental burdens that sustainable development implies will require iterative processes of reform stretching over many decades. To succeed these reforms must reflect the values and aspirations of the active elements in society. They must be backed by popular support. Thus political struggles and processes – involving the messy worlds of party politics and electoral contest – are inevitably part of the sustainable development steering problematic. At times conflicts over policy orientations will be acute. There will be continuous struggles to construct and to hold together the coalitions required to advance reform in different sectors. Here, the dynamism of communities and citizen activists, and the skills and dedication of political leaders, will be as important as any lessons from the political and administrative sciences in ensuring that governance for sustainable development remains on track.

References

Beck, U. (1994), 'The reinvention of politics: towards a theory of reflexive modernization, in reflexive Modernization: Politics, Tradition and Aesthetics in the Modern Social Order, U. Beck, A. Giddens and S. Lash (eds.).

Bhaskar, R. (1993), *Dialectic: the Pulse of Freedom*.

DAC-OECD (1993), *DAC Orientations on Participatory Development and Good Governance*, OECD.

DEFRA (2005), *Securing the Future: The UK Government Sustainable Development Strategy*, Department for Environment, Food and Rural Affairs.

Dryzek, J. (1987), *Rational Ecology: Environment and Political Economy*, Basil Blackwell.

Eckersley, R., (2004), *The Green State: Rethinking Democracy and Sovereignty*, MIT Press.

EEA (1999), Environment in the European Union at the Turn of the Century.

Giddens, A. (1984), *The Constitution of Society*, Polity Press.

Giddens, A. (1990), *The Consequences of Modernity*, Polity Press.

Glasbergen, P. (1998), (ed.), *Co-operative Environmental Governance*, Kluwer Academic, 1998.

Grin, J., F. Felix, B. Bos & S. Spoelstra (2004), 'Practices of reflexive design: lessons from a Dutch program on sustainable agriculture, International Journal of Foresight and Innovation Policy, 1: 126–149.

Habermas, J. (1996), *Between Facts and Norms: Contributions to a Discourse Theory of Law and Democracy*, translated by W. Rehg, MIT Press.

Hawken, P., A. Lovins & A. Lovins (1999), *Natural Capitalism: Creating the Next Industrial Revolution*, Little, Brown.

IPCC (2001), Third Assessment Report.

Jacobs, M. (1999), 'Sustainable development as a contested concept', in A. Dobson (ed.), *Fairness and Futurity*, Oxford University Press.

Jessop, B. (2000), 'Governance Failure' in G. Stoker. (ed.), *The New Politics of British Local Governance*, St. Martin Press, pp. 11–32.

Kemp, R., S. Parto & R. Gibson (2005), Governance for sustainable development: moving from theory to practice, *International Journal of Sustainable Development* 8: 12–30.

Kooiman, J. (2000), 'Societal governance: levels, modes and orders of social-political interaction', in J. Pierre (ed.) *Debating Governance: Authority, Steering and Democracy*, Cambridge University Press.

Kooiman, J. (2003), *Governing and Governance*, Sage.

Lafferty, W. (1996), 'The politics of sustainable development: global norms for national implementation', *Environmental Politics* 5: 185–208.

Lafferty, W. & J. Meadowcroft (2000) (eds.), *Implementing Sustainable Development: Strategies and Initiatives in High Consumption Societies*, Oxford University Press.

Luhmann, N. (1989), *Ecological Communication*, University of Chicago Press.

Marx, Karl (1852), *The Eighteenth Brumaire of Louis Bonaparte*.

McDonough, W. & M. Braungart (1998), The next industrial revolution, *The Atlantic Monthly*, October.

MEA (2006), *Ecosystems and Human Well-Being: Synthesis Report*, Millennium Ecosystem Assessment, Earthscan.

Meadowcroft, J. (1997), 'Planning for sustainable development: insights from the literatures of political science', *European Journal of Political Research* 31 (1997): 427–454.

Meadowcroft, J. (1999), Planning for sustainable development: what can be learned from the critics?', in M. Kenny and J. Meadowcroft (eds.) *Planning for Sustainability*, Routledge, pp. 12–38.

Meadowcroft, J. (2004), Deliberative democracy', in R. Durant, D. Fiorino and R. O'Leary and (eds.), *Environmental Governance Reconsidered: Challenges, Choices and Opportunities*, MIT Press, pp. 183-218.

OECD (2001), OECD Environmental Strategy for the First decade of the 21st Century.

OECD (2002a), *Governance for Sustainable Development: Five OECD Case Studies*, Paris: OECD.

OECD (2002b), *Improving policy coherence and integration for sustainable development: a checklist*, Paris: OECD.

Pierre, J. & Peters, G. (2000), *Governance Politics and the State*, Macmillian.

Plumptre, T. & J. Graham (1999), 'Governance and Good Governance: International and Aboriginal Perspectives', Institute On Governance.

Princen, T., M. Maniates & K. Conca (2002), *Confronting Consumption*, MIT Press.

Rhodes, R. (1996), The new governance: Governing without government, *Political Studies* 44, 652–666.

Schellnhuber, H. (2006), *Avoiding Dangerous Climate Change*, Cambridge University Press.

Toner, G. (2006), *Sustainable Production: Building Canadian Capacity*, UBC Press.

Torgerson, D. (1999), *The Promise of Green Politics*, Duke University Press.

UNDP (1997), 'Governance and Sustainable Human Development', United Nations Development Programme.

Voß, J. & R. Kemp (2005), 'Reflexive governance for sustainable development – incorporating feedback in social problem solving', IHDP Open Meeting in Bonn, October 9–13, 2005.

WCED (1987), *Our Common Future*, World Commission on Environment and Development,Oxford University Press.

World Bank (1991), *Managing Development: The Governance Dimension*, Washington.

WSSD (2002), 'Plan of Implementation of the World Summit on Sustainable Development', *Report of the World Summit on Sustainable Development*, Johannesburg, South Africa, 26 August–4 September 2002.

Assessing the Dutch Energy Transition Policy: How Does it Deal with Dilemmas of Managing Transitions?

RENÉ KEMP, JAN ROTMANS & DERK LOORBACH

Key Problems in Steering Societal Change Towards Sustainability Goals

Sustainable development is about the *redirection* of development (WCED, 1987). It is not about an identifiable end state (Kemp *et al.*, 2005; Meadowcroft, 1999; Meadowcroft *et al.*, 2005; Voß *et al.*, 2006). Sustainable development is a contested concept. The requirements of sustainable development are multiple and interconnected. As an inherently dynamic, indeterminate and contested concept (Mog, 2004), sustainability cannot be translated into a blueprint from which criteria can be derived and unambiguous decisions can be taken to get there. From a governance perspective such disagreement is an *essential* part of sustainable development, one that makes operationalization difficult (Farrell *et al.*, 2005).

Sustainable development sparks debate about the capacities of political steering and governance (Voss & Kemp, 2006). Capacities for steering are limited because of several problems: ambivalent goals; uncertainty about cause–effect relations and, distributed power of control (Funtowicz *et al.*, 1998; Roe, 1998; Voss *et al.*, 2006); political myopia; determination of short-term steps; and a danger of lock-in (Kemp & Loorbach, 2005).

The next section discusses these points in more detail and then uses them to evaluate Dutch energy transition policies. These issues are quite clear and well known among policy scientists. Lindblom (1959; 1965; 1979) argued for incremental politics, which is concerned with *ills* rather than visions. He later elaborated his viewpoints in a model of partisan mutual adjustment, a model that is today being called network management. The advantages of a strategy of mutual adaptation are clear: being based on *quid pro quo* it will produce mutual gains for all. It is a useful model for achieving wins through no-regret policies but ill-suited for achieving structural changes or transitions, for example in the energy system. Models of interactive governance based on negotiation will fail to achieve fundamental change *unless* there is a commitment to long-term change that is underpinned by institutions towards this end.

This article outlines a model of managing long-term change in production–consumption systems which appears to be a useful model for working towards system innovation and transitions offering sustainability benefits. The model is called transition management and is currently used by the Dutch government to manage the transition to a low-carbon energy system. Transition management combines the capacity to adapt to change with a capacity to *shape change* (Rammel *et al.*, 2004) and is concerned with positive goals (collectively chosen by society following a process of problem structuring).

Dealing with Dilemmas for Steering

The following section investigates how the mentioned problems for steering may be addressed. The problems are slightly reformulated.

Problem 1. Ambivalence About Goals

Complex societal problems related to sustainability are characterized by dissent on goals, values and means. Different people have different perspectives on what is being discussed as 'the problem', they have different values and favour different solutions. For example, there is no consensus on what sustainable energy or agriculture means in practical terms. For some people, biological agriculture is the only sustainable form of agriculture; for others the larger land requirements of biological farming make it non-sustainable in a global context. Each option has its own setbacks, some of which are only revealed in the course of time.

A proximate solution to the problem of dissent is: continuous and iterative deliberation and assessment. Even in the case of dissent about appropriate solutions, it may be possible to come to define key parameters for a future system, such as a sustainable energy system that is reliable, affordable and low in CO_2. Other parameters could be added, for instance the criterion of no biodiversity loss (relevant for bio-energy). Problem structuring methods (Rosenhead & Mingers, 2001) may be used to get to a shared problem definition about the non-sustainable aspects of the current system. This does not solve the problem of ambivalence but helps to reduce or clarify it.

Problem 2. Uncertainty About Long-Term Effects

Knowledge of ecological cause-and-effect relations is often limited, most especially in the beginning but also quite frequently at later times (e.g. acid rain processes are still not understood completely thirty years after their discovery). It took a long time to understand the effects of sulphur emissions and a very long time to link lung cancer to asbestos. We also face uncertainty about the effects of intervention and the long-term effects of socio-technical transformations, in part because the effects depend on contingencies. With regard to interventions, there is the dilemma of control (Collingridge, 1980), which holds that the capacity to shape a new development is greatest at the time when we know the least: when the effects become apparent it is difficult to alter the course of development because of sunk investments and interests vested in the continuation of the path of development.

This cautions against rapid new development(s) and calls for the creation of intelligence about long-term effects and the creation of adaptive capacity. Knowledge of the long-term system effects of technology might be understood more fully through risk assessment, technology assessment and monitoring of effects. The capacity to react can be enhanced through flexible designs (Verganti, 1999), adaptive management (Lee, 1993; Walker *et al.*, 2001), the use of portfolios and the use of capital-extensive solutions with relatively short life times (Collingridge, 1980).

Another source of structural uncertainty towards the future lies in the preferences and needs of citizens and society at large, which are subject to change. More often than not, policies are based on the presumption that societal preferences are unchangeable or at least stable. However, especially in the case of problems related to sustainability, preferences and needs will change over time and indeed also need to change. Uncertainty in this respect is thus, simultaneously, a reason for future-orientated policies to become more robust, diversified and adaptive, and a reason for developing anticipatory strategies that help influence and change current preferences and need to support a more sustainable society.

Problem 3. Distributed Control

In pluricentric societies development cannot be steered from the top (Kooiman, 1993; Pierre, 2000). Control power is distributed over various actors with different beliefs, interests and resources. Influence is exercised at different points, also within government, which consists of different layers and silos (compartments), making unitary action difficult or simply impossible. The distributed nature of power and capacities for control, calls for joint decision-making and network management (de Bruijn & Heuvelhof, 1995; Kickert *et al.*, 1997). However, current modes of network management in public governance are not equipped for long-term change. They are too little concerned with long-term ends. A form of interactive governance is needed, concerned with expressing long-term aims and the management of transition processes. The formulation of socio-technical visions for meeting long-term aspirations helps actors to coordinate their action and do useful things for their own personal sake and for society at large

(Grunwald, 2004). Visions help to make explicit what is involved in wide-ranging change, which is useful for thinking, for assessment and, of course, also for action (Smith *et al.*, 2005).

Visions fulfil positive functions for action but there are also negative aspects. First, visions are not expressed by society but by individuals or social groups (Grin, 2000). Visions are related to interests. Secondly, the outcomes of the process of change involved may not constitute true progress. There may be important societal costs that outweigh any benefits. This is why it is important to explore multiple visions and to be reflective about them. Visions for sustainability should be assessed constantly. Participatory integrated assessment is a useful approach for this.

The key question for steering here is how diverse concerns and diverse knowledge can be utilized for long-term societal change offering sustainability benefits at all levels between the local and the global. Here one could rely on possibilities for self-organization available in society. Space for interaction and innovation can be created at the national level. At the local and regional level, this can be done in the two ways: first by allowing local government an amount of discretion regarding what to do and, secondly, by stimulating local actors to reconsider basic assumptions about problems and solutions and their 'normal' way of dealing with issues—to open up the problem space and solutions space. The knowledge and judgement of local decision makers should not be suppressed but be utilized. Beck (1997) argued for the reinvention of politics, "the shaping of society from below", with actors acting on lessons learned at different places. Functional coordination, cross-linking and integration are common strategies for dealing with problems of distributed control but they should be practised reflexively.

Problem 4. Political Myopia

From historical studies (Geels, 2005), it is known that transitions in socio-technical systems take at least one generation (25 years), which means that they span various political cycles. The management of purposive transitions in some way must survive short-term political changes. There is no simple solution for this except that policy makers and politicians have to accept that a transition is needed. For politicians to accept this they have to be convinced that a problem needs fundamental change and that time is needed for such a change to occur. The change process then must be instituted gradually, through transition agendas, programmes, supporting organizations, implementation strategies and reflexive arrangements in ways that help to deal with changing circumstances and political wishes (co-evolution), which means that the institutions should be adaptive.

Problem 5. Determination of Short-Term Steps for Long-Term Change

It is often unclear how long-term change may be achieved through short-term steps. Short-term action for long-term change presents a big problem for decision makers at all levels. There exists little theory on how to do this. A dual strategy of foreseeing and backcasting based on integrated system analysis may be useful here. Such an integrated system analysis may help to analyse the characteristics and origin of a persistent problem, identify key themes of issues for change and demarcate the subject itself. The reasoning forward would be based on trend analysis and foresight, developing future visions, long-term ambitions and (intermediary) goals (Von Schomberg *et al.*, 2005; Weber, 2006). Participatory backcasting (Quist, 2007; Weaver *et al.*, 2000) and the development of transition agendas, target images and transition paths (Loorbach 2007; Rotmans, 2005) help to identify strategic experiments and help to set goals for new sociotechnical systems.[1] Such a process helps to identify useful steps in the form of short-term actions that generate useful lessons and facilitate further change (Grin & Weterings, 2005; Kemp & Loorbach, 2005). An important role is envisaged for strategic experiments for learning about user satisfaction, sustainability aspects, critical problems and for creating networks for cooperation.

Problem 6. Danger of Lock-in

As with old solutions, there is a danger that one gets locked into particular solutions that can be viewed as 'non-optimal' from a longer-term perspective.[2] A solution for this is the development and use of a portfolio of options (Weber, 2006) in the context of a transition agenda. A transition agenda is a shared strategy for social change that is based on a shared consensus about the need for structural change and about an overall direction. Within this context, so-called thematic target images are brought together with a collection of different transition paths towards these images. This approach allows for diversity and competition while simultaneously maximizing the potential for synergies and co-evolution. At the level of particular solutions (i.e. the level of options within transition paths) there is a lot of uncertainty about which solution (i.e. technology option, investment) is best. Portfolio management is a good strategy here which is widely practiced in finance and large business. Support for options could be based on promises and specific benefits for the nation or region in which it is used. Support should be given to not just one option but to a portfolio. The portfolio of promising solutions would safeguard against the operation of markets favouring short-term solutions. Of course, the portfolio should be regularly reviewed and adapted, i.e. it should be dynamic. Phase out of support should be part of portfolio management. A second strategy is simply to be prudent, and not to go for the best available solution but to take time to wait for better solutions and spend money on their creation.

   All this serves to show that at least proximate solutions (ideas for what to do) exist for the six identified problems for steering, which in our view constitute the most important problems for sustainable development policy.

The Model of Transition Management

The model of transition management is a model for governance developed to deal with persistent problems that require systemic change. Persistent problems are complex, uncertain, difficult to manage, hard to grasp and operate at different scale levels (Rotmans, 2005). Examples are the global climate change problem, the agricultural problem and the mobility problem. They are deeply rooted in our societal structures and there are no ready-made solutions for them: sectoral or partial solutions already soon become part of the persistent nature of these problems. The model of transition management has been developed by the authors, in interaction with policy makers, and is described in Rotmans *et al.* (2000; 2001) and elaborated in Kemp & Rotmans (2004; 2005), Kemp *et al.* (2007) and Loorbach (2007). The basic philosophy is *goal-orientated modulation*: the utilization of ongoing developments for societal goals. The model is based on notions from Integrated Assessment and insights from innovation theory, especially the work on technological transitions (Freeman & Perez, 1988; Geels, 2002; 2005), the work on path-dependence (Arthur, 1989; David, 1985) and on sustainable development and participatory policy approaches (Rotmans, 1998).

Transition management is a new steering concept that relies on 'darwinististic' processes of guided variation and selection instead of planning. Collective choices are made 'along the way' on the basis of (new) learning experiences at different levels. Different trajectories are explored and flexibility is maintained, which is exactly what managers would do when faced with great uncertainty and complexity: instead of defining end states for development, they set out in a certain direction and are careful to avoid premature choices.

Key elements of the transition management cycle are anticipation, learning and adaptation. The starting point is the structuring of problems. This is followed by the development of long-term visions and goals. Visions of sustainable development for energy supply and other domains are being explored through transition experiments as part of programmes for system innovation that are defined in transition arenas, bringing together private and public actors. Transition management as a model of governance relies on a cycle of problem structuring, visioning, experimentation, policy development, implementation and adaptation.

The visions help to define experiments and programmes for system innovation, the lessons of which should lead to a *revision* of the visions and to the identification of new things to do (new experiments and changes in the policy framework).

Transition management has elements of planning through the use of goals and programmes for system innovation but does not aim to control the future. It relies heavily on market forces and decentralized decision making. It does not blankly rely on market forces, but is concerned with the conditions under which market forces operate, by engaging in 'context control' so as to orientate market dynamics towards societal goals. It consists of government acting to secure circumstances that will maximize the possibilities for progressive social movement by promoting innovation and mitigating negative effects (Meadowcroft, 1999). Private initiative is thus not curtailed but rather reorientated towards those activities that serve not only private goals but also serve social goals. This is done through programmes for system innovation and through the use of policy goals providing guidance to societal actors.

Conflict is kept within bounds but is accepted and even viewed necessary (same as in the "compass and gyroscope" model of Lee (1993) for combining science with politics). The structuring form is that of *heterarchy*: modification of structural links and modification of the self-understanding of actors (identities), strategic capacities and interests of individuals and collective actors and hence their preferred strategies and tactics (Jessop, 1997).

Transition Policy for Sustainable Energy in the Netherlands

This section examines the ways in which the problems of steering were confronted with transition management as a guide. A full account of Dutch transition policies cannot be given; refer to Harmsen (2006), Kemp & Loorbach (2005), Kern & Smith (2007) and Rotmans (2005) for that. The model of transition management informed Dutch transition policies as a general frame (ideograph). It was originally developed for the Dutch government in 2000 who embraced the transition concept in the government white paper *A willing world* [*Een wereld en een wil*]. The government took over a good deal of the concept of transitions and transition management. It took a long-term orientation as a starting point and adopted a strategy of keeping options open for a certain time. It created transition arenas (platforms), developed transition pathways and started up about one hundred experiments. In doing so, it adhered to the principles of transition management. By 2007 the new cabinet has taken the energy transition as one of its pillars to achieve a sustainable energy supply system in the Netherlands. But it started as a niche project within the Ministry of VROM (Environment and Spatial Planning) and was picked up as a serious policy experiment by the Ministry of Economic Affairs. This section assesses in what way the Ministry of Economic Affairs has dealt with the problems of steering towards sustainable development by using transition management.

Problem 1. Dissent and Ambivalence About Goals

There was little dissent about the need for change amongst the actors involved in sustainable energy (ministries, business companies, NGOs and knowledge institutions), but a broad public and political awareness regarding the issue was absent around 2000. Because of global changes, such as the Middle East conflict, rising oil prices, Russia's gas threat, but also local climate pollution, high energy bills and extreme weather, this has changed significantly. It helped to create an image of the current energy system being unsustainable and increased the attention given to nuclear power, CO_2 storage and biomass import as possible solutions. There is by now a wide consensus among the energy experts that the current energy system based on fossil fuels is not sustainable environmentally, socially or economically. The Netherlands was and still is committed to greenhouse gas reductions through the Kyoto process which is viewed as a first step. Fossil fuel-based visions based on carbon capturing and sequestering are accepted so far, in addition to other visions based on renewables and energy efficiency. Nuclear energy was put forward as a transitional (temporary) option by the energy transition task force but did not become an official transition path. Neither the Ministry of Economic Affairs nor the industrial actors active in the platforms of the energy transition favoured it, but there seems to be a strong national and international lobby. Dissent did not obstruct the selection of transition paths and setting of long-term goals, mainly because this approach allows for diversity and different, competing, options based on the idea of variation and selection. This portfolio approach fitted with Dutch culture and growing policy beliefs that it is best to rely on an adaptive portfolio. The exploration and use of various options fits with the model of transition management, which is based on an evolving portfolio.

The vision that ultimately emerged from this process, a sustainable energy system in 2050 is defined as: (i) clean (climate-robust, i.e. a CO_2 reduction of 50 per cent); (ii) affordable (low prices and functional); and (iii) secure (guaranteed and reliable supplies). This is supplemented by more specific goals such as the goal to increase the annual rate of energy saving from 1.5 per cent to 2 per cent a year and the target of 30 per cent for green energy sources by 2030. The goals were set by the energy transition platforms created by the Ministry of Economic Affairs.

Problem 2. Dealing with Uncertainty

The long-term system effects of various energy systems were not given much attention. They were explored only superficially. A long-term study of ECN about energy and society in 2050 looked at the various long-term options but did not examine in any detail the sustainability aspects.

A scenario analysis was used to identify robust elements of energy futures, leading to the selection of new gas, chain efficiency, biomass resources (for energy but also for other purposes), alternative motor fuels, and sustainable electricity as main routes (*hoofdroutes*) of the energy transition. During the process scenario thinking and other approaches for exploration were not used systematically at the level of the whole energy system in international context, although a number of platforms also used forecasts and predictions specific for their subject. For the main routes special transition paths were selected by specially created transition platforms involving private and public actors. The choice of

Table 1. The energy transition themes, goals and paths chosen by the platforms

Theme	Goal	Transition path
New gas	To become the most sustainable gas country in europe	Decentralized electricity generation Energy efficient greenhouses Green gas hydrogen Clean fossil fuels
Sustainable mobility	Factor 2 reduction of GHG emissions for new vehicles in 2015 and factor 3 reduction for all vehicles in 2030	Hybrid propulsion Biofuels Hydrogen vehicles Intelligent transport systems
Green resources	Substitution of 30% of resources for energy by green resources by 2030	Biomass production in NL Chains for biomass import WISE Biomass co-production Synthetic Natural Gas Sustainable chemistry
Chain efficiency	20–30% extra improvement of product chains by 2030	Optimising the waste chain Precision farming Process intensification Multimodal transport Clearing house for bulk products Symbiosis (closing material loops) Micro cogeneration Energy efficient paper production
Sustainable electricity supply	To make electricity supply more sustainable	Renewable energy sources Decarbonisation and cogeneration Electric infrastructure Electricity use
Built environment	To accelerate energy improvement programmes and stimulate new innovations	Energy improvements in built environment Development and implementation of innovations Removal of institutional barriers

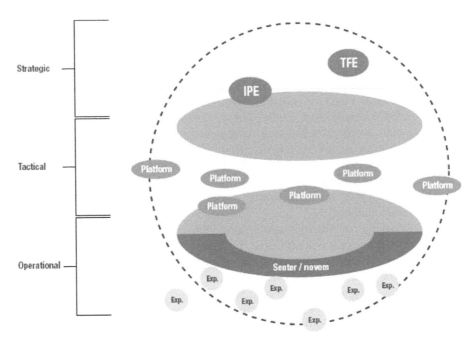

Source: Loorbach (2007)

Note: Exp. stands for transition experiment; SenterNovem is the government agency responsible for implementation of innovation policy

Figure 1. The strategic, tactical and operational elements of transition management

the transition paths occurred over a period of six years and resulted in the selection of 28 official transition paths. An overview of the paths is provided in Table 1.

The paths are paths for exploration, not implementation. The crux of dealing with uncertainties is that, rather than making definite choices, small-scale experiments are set up and executed from which much can be learned, so that better information is available later on the (un)sustainable aspects of pathways and the related experiments. In this respect better-defended choices can be made by better-informed actors, such as decision makers. Some paths will obtain extra support (from public and private sources) than others.

Problem 3. Distributed Control

The issue of distributed control was dealt with through the creation of a task force (TFE) in 2005 and the creation of a directorate for policy coordination across various ministries (IPE), the creation of networks and platforms of private and public actors to identify attractive transition visions, paths and, finally, the fostering of coalitions for transition experiments. In theoretical terms these are referred to as strategic, tactical and operational types of transition management (Loorbach, 2007). A graphical representation of this can be seen in Figure 1.

The transition platforms form the heart of the transition (Aubert, interviewed in Kern & Smith, 2007) and their activities are described further on. They played a pivotal role in the selection of main routes, the selection of transition paths and identification of transition experiments and the development of the broader transition community. Responsibilities for the selection of transition paths were devolved to the transition platforms, bringing together business actors, energy experts, people from government and civil society.

At this tactical level various networks, alliances and communities were formed around the diverging transition pathways. New institutional arrangements were developed, amongst others the UKR-arrangement (unique chance arrangement), and a service office for frontrunners.

In May 2006 a transition action plan was presented to the Dutch government. The action plan, written by the task force energy transition, was based on inputs from the platforms. Apart from presenting 26 paths, it argued for the doubling of energy innovation expenditures (from €1 billion to €2 billion a year to be paid out of the general government budget instead of through special taxes) and made a plea for "consistency and continuity of policy based on a long-term vision about sustainable energy".

For discussing transition issues among various departments and to foster collaboration, a directorate was created: the interdepartmental programme directorate energy transition (IPE). The directorate is located at the Ministry of Economic Affairs and encompasses 30 civil servants from six ministries. The impulse for the directorate came from stakeholders involved in the energy transition who developed pressure on the government to re-organize policies and combine them (Kern & Smith, 2007). The creation of such a directorate fits with the transition management philosophy and is an example of endogenous institutionalization (self-organization).

Problem 4. Political Myopia

Structural change in infrastructure-bound systems, such as the energy system, is politically difficult. Any fundamental change in an industry's core technologies creating losers, such as regional unemployment problems, is difficult to bring about politically and is possible only exceptionally (Janicke & Jacob (2005), quoted in Kern & Smith, 2007). Substantial political stability and resilient coalitions are required to keep reform from being derailed (Meadowcroft *et al.*, 2005). So far, transition management has survived four government changes (new cabinet Kok and three cabinets Balkenende). The reason for this is that in the first five years it was, politically, not very salient; transition matters were in the safe hands of ministries. The report of the task force in May 2006 changed this by attracting more attention to the energy transition. What began as an experiment in policy innovation became an institutionalized process, which is currently supported by the main actors: Ministries, cabinet and the main advisory councils—SER and VROMRaad. Political parties accept the basic idea of an energy transition. Arguably, the ambiguity of transition ideas aided their popularity (Hendriks, 2006, Smith & Kern, 2007).

The problem of distributed powers of control is thus dealt with by giving an important role to the transition platforms. In later stages the role of the transition platforms will become less important, and politics more important. Firm support has been obtained for the energy transition, with many parties collaborating with each other. The problem of distributed control was thus 'solved' through the strategic use of distributed control. Without the binding notion of transition, not so many parties could have been mobilized.

Problem 5. Determination of Short-Term Steps for Long-Term Change

Perhaps the strongest aspect of transition management is that it offers a practical model for connecting long-term thinking with short-term action without relying on planning. The selection of transition paths and the use of strategic experiments, called transition experiments, help to work towards long-term change in the form of system innovation. The transition experiments (old and new ones) should set into motion learning processes and foster institutional change, such as the creation of networks, standards, new procedures and practices. They constitute a bottom-up mechanism, which should work in tandem with top-down activities consisting of choices about transition paths eligible for support and government policies.

Transition experiments are supported by the "Unieke Kansen Regeling" of €35 million.[3] In order to qualify for support the experiments should: (i) be part of an official transition path; (ii) involve stakeholders in an important way; and (iii) have explicit learning goals for each of the actors of the consortium. Transition experiments for new gas are:

- buses on natural gas in Haarlem/Rijnmond;
- liquefied natural gas as a substitute for diesel;
- CO_2 delivery to greenhouses in the horticulture sector;
- urban transport using compressed natural gas in the north of the Netherlands;
- heating from biogas in the Polder district in Zeewolde;
- pilot project on micro-cogeneration in households.

Problem 6. Danger of Lock-in

The issue of lock-in received a lot of attention in transition thinking, both the current lock-in to fossil fuels and dangers of becoming locked into new solutions. It is one of the reasons why a dynamic portfolio of options is supported. It is worth noting that the director of the energy transition (also the chairman of the IPE), Hugo Brouwer, perceived the energy transition in evolutionary terms. He sees the energy transition as a process of "accelerated evolution" (Shell-venster, May/June 2005), i.e. the outcome of processes of variation, selection and retention. Selection would occur through the market and political process. The current portfolio is very broad. Maybe it is too broad but it will become narrower in the process of market introduction. It is interesting to observe that the task force in the action plan for energy transition argued for a new coal-gasification plant

Table 2. Participation in the private-public platforms of the energy transition (Kern & Smith, 2007)

Platform	Government	Business	NGOs	Intermediaries[1]	Science	Total
Green Resources	1	6	1	1	6	15
New Gas	1	6	1	1	3	12
Chain Efficiency	1	6	0	1	3	11
Sustainable Mobility	3	10	3	0	0	16
Sustainable Electricity	1	3	0	0	3	7
Built Environment	0	4	4	2	1	11

[1] The category *Intermediaries* encompasses representatives from municipalities, SenterNovem (excluding the secretaries), the provinces, regional initiatives (such as Rijnmond) or national advisory boards such as SER.

and a new nuclear plant for achieving quick CO_2 reductions, to complement the transition paths approach orientated towards system innovation. These goals, along with the *modus operandi* of the task force, were not received well by the platforms. Whilst the portfolio is broad, the criteria on which it is based are rather narrow. There was a very strong focus on CO_2 reduction.

Differences with the TM Model

In what ways do the transition initiatives differ from the model of transition management? The first and main difference is that outsiders are scarcely involved (Hendriks, 2006; Hofman, 2006; Loorbach, 2007; Loorbach & Kemp 2007; Rotmans, 2005). The process is dominated by regime actors (business and energy specialists), as can be seen in Table 2. Regime actors play an important role, especially in the task force headed by Rein Willems, the director of Shell-Nederland.

Hendriks (2006) spoke of "a democratic disconnect". Whilst she is right on this, several actors invited to the platforms declined to be involved. The *Consumentenbond*, the largest consumer organization, felt it lacked knowledge about energy innovation and transition issues. The projects themselves are innovative and some of the transition paths are very radical also for the companies concerned. None the less it would have been better if more niche parties could have been involved as one can expect them to embrace radical innovation, since they have no interest in the *status quo* (Rotmans, 2005).

Secondly, up until now, demand-side issues and wider issues of societal embedding have been neglected. The transition experiments are very technological by nature; they are hardly aimed at institutional or cultural change. They consist of rather low-risk projects primarily related to CO_2 reduction (and not, for instance, to security). Further, the aspect of scaling up experiments is not addressed at all: how the experiment may foster changes in networks, mental models, structures and regimes and, in so doing, contribute to a regime shift.

Thirdly, little attention was given to strategic issues of integrated system analysis and problem structuring. An old scenario for the energy system was used. Participatory scenario development (advocated for transition management by Sondeijker et al., 2006) was not part of the process (Loorbach & Kemp, 2007). Sustainability assessment did not play an important role. Only for biofuels was a large study commissioned to determine criteria for "sustainable biofuels".

Fourthly, the role of the task force is dubious from a transition perspective. On the one hand it has led to an acceleration of the energy transition process, because they set the energy transition high on the political agenda, influencing high-level politicians and policy makers. However, on the other hand, the current task force is quite autonomous and imposing, putting pressure on the other bodies/constituencies of the transition process. It consists mainly of regime players, trying to defend their own interests, while the very idea behind transition process is to create enough space for frontrunners, pioneers (first movers). The task force might limit the amount of space for frontrunners.

The model of transition management has thus been followed only partially. It is not believed that the above issues will be taken up in some way. The platforms have already been asked to pay more attention to non-technology issues. As a possible sign, all three transition paths of the sixth platform founded in 2007 are non-technological. They are about creating institutions and incentives for energy saving. Also, the role and composition of the task force and platforms are being debated, probably leading to significant changes in the near future. This is aided by the central position of energy transition in the new cabinet's policies, which has led to debates and political games within the government over responsibilities, money and stakes.

Overall, energy transition in the Netherlands is considered a success in creating a new discourse, framework and orientation which is widely supported. Nevertheless, although many of the recommendations of the original model have been followed, transition management is not the open, reflexive process it was supposed to be. The transition paths have been chosen by people in the platforms (in which the business voice is prominent). There has been little cooperation between the platforms or mutual learning. It has not become politically salient in parliament and society is not really involved in it. The portfolio of alternative energy technologies is very broad but this is not necessarily a bad thing. There is a danger that by formalizing the transition paths as a basis for investment, regulation and policy decisions, they become a goal in themselves instead of a means. On the other hand, the issue of lock-in is being considered. It is one of the reasons why the Ministry of Economic Affairs opts for a portfolio approach and relies on the use of variation and selection processes.

So far the attention to transitions has not resulted in the changes in fiscal policies or in environmental policies that will be needed to change the energy supply system (Berkhout *et al.*, 2004).[4] Instrumental issues about transition management are currently worked out by the task force and people from the ministries. It will be interesting to see what kind of market pull is going to be used and how the government or a new transition council is going to manage the portfolio of technology options. Public procurement will probably be one of the instruments that will be used, together with subsidies and long-term agreements. The government wants to achieve a good balance between deployment and innovation. Cost efficiency is an important consideration for deployment policies. The creation of new business opportunities is a criterion for innovation. For the built environment, regulations will be used. The regulations will not be technology prescriptive. No plans are being made to phase out unsustainable energy technologies, which suggests that a regime change has to occur through markets.

Reflections

Sustainable development is generally viewed as requiring a change in the trajectories of development (WCED, 1987; Weaver *et al.*; 2000) but we lack models to achieve this. Any model of sustainable development must deal with problems of ambiguity and dissent, uncertainty and distributed powers of control. This article described a model of reflexive governance that aims to modulate ongoing developments to sustainability goals through changes in governance (participatory and value-focused) and adaptive policies for system change. This model is called transition management for the reason that it is concerned with transition processes of systems change. The model is currently used in the Netherlands to 'manage' transitions to alternative energy, agriculture and transport systems. It is believed to be an interesting model for sustainable development because it deals with the problems of ambiguity of goals, uncertainty about socio-economic dynamics, distributed control and political myopia through reflexive learning (using integrated assessment, problem-structuring and social experiments).

Transition management is a new steering concept that relies on processes of variation and selection by making use of 'bottom-up' developments but also top-down elements, such as long-term goals (based on a sustainability vision) both at the national and local level. Governing processes are opened up for inter-action and feedback relations with an important role for subpolitics (Beck, 1994; 1997) and actors interested in system innovation.[5]

The use of transition management for the transition to a sustainable energy system illustrates how collective long-term ambitions and a shared agenda can go hand-in-hand with short-term diversity, experiments and even dissent. Tran-sition management is, by definition, about using the energy that arises out of the interaction between long-term consent and short-term dissent through learn-ing-by-doing and doing-by-learning. By an integrative and outward-looking analysis of societal dynamics, the capacity to anticipate is being improved and alternatives can be developed in a timely fashion.

Choices about goals, means and instruments are being made as part of itera-tive processes, although the final decisions and implementation are still predomi-nantly made through formal and regular political and bureaucratic structures. It is not a return to planning but a form of context steering that is concerned explicitly with learning and innovation, including institutional innovation.

The principles of transition management have been followed broadly, but now a critical phase occurs when important instrument choices are to be made. Much of the success of transition management will depend on these. The policy commitment to achieving a transition in the energy system and creation of networks of innovative actors should help to make instrument choices (such as fiscal changes, special funds, the use of regulations and use of innovation waivers) necessary for a transition to occur. Transition management should thus help to do what economists tell govern-ments to do: to internalize external costs—something that is hard to do in a world of special interests. By creating special interests in system innovation it may be easier to take those measures and to go beyond support measures in the form of subsidies.

Whether a radical change in the energy system will be achieved through tran-sition management remains to be seen. However, it has already been shown that problems of ambiguity, uncertainty and distributed control need not necessarily paralyse society and that it is possible to conceptualize and implement an innova-tive mode of governance in which the best of two worlds—planning and incre-mentalism—are brought together.

Acknowledgement

The authors would like to thank Jan-Peter Voß, John Grin, Jos Timmermans and two referees for extensive comments and suggestions.

Notes

1. This is analogous to road mapping, a business tool for exploring paths for development and iden-tifying business steps towards long-term business change (Phaal *et al.*, 2003).
2. It is impossible to establish whether something that 'could have been' is better than what we have. Non-optimality refers to a widely shared view that a practice or system is not the best possible thing.
3. So far 25 million euro has been spent: €10 million in 2005 and €15 million in 2006.
4. Biofuels are exempt from taxes but this was motivated very much by the EU Directive on biofuels.
5. In the Dutch energy transition, the 'subpolitics' consist of the delegation of transition choices to the transition platforms composed of private and public actors.

References

Arthur, W. B. (1989) Competing Technologies, Increasing Returns, and Lock-in By Historical Events, *Economic Journal*, 99, pp. 116–131.

Beck, U. (1994) The Reinvention of Politics: Towards a Theory of Reflexive Modernization, in: U. Beck, A. Giddens & S. Lash (Eds) *Reflexive Modernization*, pp. 1–55 (Cambridge: Polity Press).

Beck, U. (1997) *The Reinvention of Politics* (Cambridge: Polity Press).

Berkhout, F., Smith, A. & Stirling, A. (2004) Socio-technological Regimes and Transition Contexts, in: B. Elzen, F. W. Geels & K. Green (Eds) *System Innovation and the Transition to Sustainability. Theory, Evidence and Policy*, pp. 48–75 (Cheltenham: Edward Elgar).

de Bruijn, J. A. & Heuvelhof, E. F. (1995) *Netwerkmanagement: strategieën, instrumenten, normen* (Utrecht: Lemma).

Collingridge, D. (1980) *The Social Control of Technology* (New York: St. Martin's Press).

David, P. (1985) Clio and the Economics of Qwerty, *American Economic Review*, 75(2), pp. 332–337.

Farrell, K., Kemp, R., Hinterberger, F., Rammel, C. & Ziegler, R. (2005) From *for* to Governance for Sustainable Development in Europe—what is at stake for further research, *International Journal of Sustainable Development*, 8(1/2), pp. 127–150.

Freeman, C. & Perez, C. (1998) Structural Crises of Adjustment, Business Cycles and Investment Behaviour, in: G. Dos, C. Freeman, R. Nelson, G. Silverberg and L. Soete (eds), *Technical Change and Economic Theory*, London, Pinter Publishers, pp. 38–66.

Funtowicz, S., Ravetz, J. R. & O'Connor, M. (1998) Challenges in the use of science for sustainable development, *International Journal of Sustainable Development*, 1(1), pp. 99–107.

Geels, F. W. (2002) Technological transitions as evolutionary reconfiguration processes: A multi-level perspective and a case-study, *Research Policy*, 31(8/9), pp. 1257–1274.

Geels, F. W. (2005) *Technological Transitions and System Innovation. Co-Evolutionary and Socio–Technical Analysis* (Cheltenham: Edward Elgar).

Grin, J. (2000) Vision assessment to support shaping 21st century society? Technology assessment as a tool for political judgement, in: J. Grin & A. Grunwald (Eds) *Vision Assessment: Shaping Technology in 21st Century Society. Towards a Repertoire for Technology Assessment*, pp. 9–30 (Heidelberg: Springer Verlag).

Grin, J. & Weterings, R. (2005) Reflexive monitoring of systems innovative projects: strategic nature and relevant competences. Paper presented at the 6th Open Meeting of the Human Dimensions of Global Environmental Change Research Community Conference, Bonn, Germany.

Grunwald, A. (2004) Strategic knowledge for sustainable development: the need for reflexivity and learning at the interface between science and society, *International Journal of Foresight and Innovation Policy*, 1(1/2), pp. 150–167.

Harmsen, R. (2006) Transitiemanagement, zonde van de energie? Onderzoek naar de (on)mogelijkheden van het initiëren en sturen van een energietransitie [Transition management, a waste of energy?. A study into the possibilities to initiate and steer a transition in energy], Graduation Thesis, University of Utrecht.

Hendriks, C. M. (2006) On inclusion and network governance. The democratic disconnect of Dutch energy transitions, *Public Administration*, in press.

Hofman, P. (2006) Evaluating Energy Transition Policy. Paper presented at SWOME marktdag, Den Haag, 12 October.

Jessop, B. (1997) The Governance of Complexity and the Complexity of Governance: Preliminary remarks on some problems and limits of economic guidance, in: A. Amin & J. Hausner (Eds) *Beyond market and Hierarchy. Interactive Governance and Social Complexity*, pp. 95–128 (Cheltenham: Edward Elgar).

Kemp, R. & Loorbach, D. (2005) Dutch Policies to Manage the Transition to Sustainable Energy, in: *Jahrbuch Ökologische Ökonomik 4 Innovationen und Nachhaltigkeit*, pp. 123–150 (Marburg: MetropolisVerlag).

Kemp, R., Loorbach, D. & Rotmans, J. (2007) Transition management as a model for managing processes of co-evolution, *The International Journal of Sustainable Development and World Ecology* (special issue on (co)-evolutionary approach to sustainable development), 14, pp. 78–91.

Kemp, R., Parto, S. & Gibson, R. B. (2005) Governance for Sustainable Development: Moving from theory to practice, *International Journal of Sustainable Development*, 8(1/2), pp. 13–30.

Kemp, R. & Rotmans, J. (2004) Managing the Transition to Sustainable Mobility, in: E. Boelie, F. W. Geels & K. Green (Eds) *System Innovation and the Transition to Sustainability: Theory, Evidence and Policy*, pp. 137–167 (Cheltenham: Edward Elgar).

Kemp, R. & Rotmans, J. (2005) The management of the co-evolution of technical, environmental and social systems, in: M. Weber & J. Hemmelskamp (Eds) *Towards Environmental Innovation Systems*, pp. 33–55 (Heidelberg/New York: Springer Verlag).

Kern, F. & Smith, A. (2007) Restructuring energy systems for Sustainability? Energy transition policy in the Netherlands, Sussex Energy Group, SPRU, University of Sussex.

Kickert, W. J. M., Klijn, E. H. & Koppenjan, J. F. M. (1997) *Managing Complex Networks. Strategies for the Public Sector* (London: Sage).

Kooiman, J. (Ed.) (1993) *Modern Governance. New Government–Society Interactions* (London: Sage).

Lee, K. N. (1993) *Compass and Gyroscope. Integrating Science and Politics for the Environment* (Washington D.C.: Island Press).

Lindblom, C. E. (1959) The Science of Muddling Through, *Public Administration Review*, 19, pp. 79–99.

Lindblom, C. E. (1965) *The Intelligence of Democracy* (New York: Free Press).

Lindblom, C. E. (1979) Still Muddling, Not Yet Through, *Public Administration Review*, 39, pp. 517–526.

Loorbach, D. (2007), *Transition management: new mode of governance for sustainable development* (Utrecht: International Books).

Loorbach, D. & Kemp, R. (2007) Transition management for the Dutch energy transition: the multilevel governance aspects, in: J. C. J. M. van den Bergh & F. R Bruinsma (Eds) *The Transition to Renewable Energy: Theory and Practice*, (in press) (Cheltenham: Edward Elgar).

Meadowcroft, J. (1999) Planning for Sustainable Development: What can be learned from the critics?, in M. Kenny & J. Meadowcroft (Eds) *Planning sustainability*, pp. 12–38 (London and NY: Routledge).

Meadowcroft, J., Farrell, K. N. & Spangenberg, J. (2005) Developing a framework for sustainability governance in the European Union, *International Journal of Sustainable Development*, 8(1/2), pp. 3–11.

Mog, J. M. (2004) Struggling with Sustainability—A Comparative Framework for Evaluating Sustainable Development Programs, *World Development*, 32(12), pp. 2139–2160.

Phaal, R., Farrukh, C. & Probert, D. (2003) Technology Roadmapping: linking technology resources into business planning, *International Journal of Technology Management*, 26(1), pp. 2–19.

Pierre, J. (2000) *Debating Governance: Authority, Steering, and Democracy* (USA: Oxford University Press).

Quist, J. (2007) *Backcasting for a sustainable future; the impact after 10 years* (Delft: Eburon).

Rammel, C., Hinterberger, F. & Bechthold, U. (2004) Governing Sustainable Development—A Co-evolutionary Perspective on Transitions and Change, GoSD working paper 1.

Roe, E. (1998) *Taking Complexity Seriously. Policy Analysis, Triangulation and Sustainable Development* (Boston: Kluwer Academic Publishers).

Rosenhead, J. & Mingers, J. (2001) *Rational Analysis for a Problematic World Revisited: problem structuring methods for complexity, uncertainty and conflict* (Chicester: Wiley).

Rotmans, J. (1998) Methods for IA: The challenges and opportunities ahead, *Environmental Modeling and Assessment*, 3(3), pp. 155–179.

Rotmans, J. (2005) Societal innovation: between dream and reality lies complexity. Inaugural address at Erasmus University Rotterdam, Rotterdam.

Rotmans, J., Kemp, R. & van Asselt, M. (2001) More Evolution than Revolution. Transition Management in Public Policy, *Foresight*, 3(1), pp. 15–31.

Rotmans, J., Kemp, R., van Asselt, M., Geels, F., Verbong, G. & Molendijk, K. (2000) Transities & Transitiemanagement. De casus van een emissiearme energievoorziening. Final report of study "Transitions and Transition management" for the 4th National Environmental Policy Plan (NMP-4) of the Netherlands, October 2000, ICIS & MERIT, Maastricht.

Smith, A. & Kern, F. (2007) The Transitions Discourse in the Ecological Modernisation of the Netherlands, SPRU electronic working paper series no. 160.

Smith, A., Stirling, A. & Berkhout, F. (2005) Governance of Sustainable Sociotechnical Transitions, *Research Policy*, 34, pp. 491–150.

Sondeijker, S., Geurts, J., Rotmans, J. & Tukker, A. (2006) Imagining sustainability: the added value of transition scenarios in transition management, *Foresight*, 8(5), pp. 15–30.

Verganti, R. (1999) Planned flexibility. Linking Anticipation and Reaction in Product Development Projects, *Journal of Product Innovation Management*, 16(4), pp. 363–376.

Von Schomberg, R., Guimarães Pereira, A. & Funtowicz, S. (2005) Deliberating Foresight-Knowledge for Policy and Foresight-Knowledge Assessment, European Commission, DG Research. Available at ftp://ftp.cordis.europa.eu/pub/foresight/docs/deliberating-foresight.pdf (accessed 5 October 2007).

Voß, J.-P., & Kemp, R. (2005) Reflexive Governance: Learning to cope with fundamental limitations in steering substainable development, *Futures*, 00, pp. 00–01.

Voss, J.-P., & Kemp, R. (2006) Sustainability and Reflexive Governance: Introduction, in: J.-P. Voss, D. Bavknecht & R. Kemp (eds) Reflexive Governance for Sustainable Development, pp. 3–28. (Edward Elgar: Cheltenham).

Walker, W. E., Rahman, A. S. & Cave, J. (2001) Adaptive policies, policy analysis, and policy-making, *European Journal of Operational Research*, 128, pp. 282–289.

WCED (Ed.) (1987) *Our Common Future*. Final Report of World Commission on Environment and Development (New York: WCED).

Weaver, P., Jansen, L., van Grootveld, G., van Spiegel, E. & Vergragt, P. (2000) *Sustainable Technology Development* (Sheffield: Green Publishing).

Weber, K. M. (2006) Foresight and adaptive planning as complementary elements in anticipatory policy-making: a conceptual and methodological approach, in: J-P. Voß, D. Bauknecht & R. Kemp (Eds) *Reflexive Governance for Sustainable Development*, pp. 189–221 (Cheltenham: Edward Elgar).

Contextualizing Reflexive Governance: the Politics of Dutch Transitions to Sustainability

CAROLYN M. HENDRIKS & JOHN GRIN

Introduction

Steering towards sustainability takes place in a dynamic political world. Nothing is static. Dynamism is, of course, an inherent feature of politics, but in contemporary liberal democracies we are witnessing change on many fronts. Actors and institutions are diversifying, rules and values are shifting, and political, market and social forces are evolving (Alexander, 1995; Kooiman, 1993). With these changes we have seen greater interdependencies between actors and institutions, as well as more uncertainties and ambiguities (Pierre, 2001; Rhodes, 1997).[1] Our starting point here is that these conditions are relevant not only to environmental matters but are symptoms of ongoing political modernization (Arts & van Tatenhove, 2004; Hajer, 2003). The question considered empirically in this contribution is: how does steering for sustainability work within this changing governance context?

As we understand it, governing for sustainability requires a continuous reconsideration of the practices, structures and outcomes of governance. This means moving on from the (weak) 'ecological modernization' agenda of the 1990s and rethinking not only how we *arrange* our social-technical systems, but also how we *govern* them.[2] In other words, steering for sustainability involves an ongoing and fundamental review of the governance system itself; the institutions of the state, market, civil society and science, as well as their mutual alignment (Grin, 2006, pp. 57–61). Reflexive governance of this kind has several significant implications for actors and institutions. First, it demands that they reflect on how their frames, structures and patterns of action contribute to persistent problems (Voß & Kemp, 2005). Actors are also encouraged to *loosen their grip* on the desire 'to control' problems in the way that classical modernity prescribes (Beck *et al.*, 2003; Grin, 2006). Thirdly, reflexive governance requires innovative and strategic thinking to fundamentally transform existing practices and structures (Loorbach & Rotmans, 2006; Schot, 1998).

This paper explores how this radical and reflexive mode of steering works in practice alongside a dynamic governance landscape. In order to explore this question empirically we start from the perspective that, in practice, steering for sustainability typically surfaces as isolated moments of reflexivity amid a sea of everyday politics.[3] We refer to these moments as *reflexive arrangements*. These might be one-off projects or a series of programmes in which actors work together to scrutinize and reconsider existing systems, and the broader rules and paradigms within which they operate (see Grin *et al.*, 2004).

The empirical focus of our contribution is on how *reflexive arrangements* interface with their broader socio-political context. To this end, a conceptual framework is developed first, which links reflexive arrangements to relevant arenas of political and public debate. This is then applied to a Dutch case study on agricultural reform. Our analysis examines how actors engage in a process of steering for sustainability in a context where institutions and actors are multiple and power fractured. We pay particular attention to how sites of steering are negotiated and used by various actors to legitimize their actions and roles. Before discussing the empirical case we survey and critique emerging literature on reflexive governance for sustainability and call for a more politically situated conceptualization of its practice.

Conceptualizing Reflexive Governance for Sustainability

Reflexive governance is not an easy term to pin down. Its ambiguity stems largely from the multiple faces of reflexivity (see Lynch, 2000).[4] In its most elementary meaning, 'to be reflexive' is the capacity to turn or bend back on oneself. On the political stage, reflexivity can be enacted in different ways (e.g. Berejikian & Dryzek, 2000; Considine, 2000; Cunliffe & Jun, 2005). For those working with sustainability in mind, reflexivity has come to take on a particular meaning, one that is best appreciated by Voß & Kemp's (2005) distinction between first- and second-order reflexivity. Under this schema, **first-order reflexivity** refers to the continuous cycle of side effects from simple modernity. Reflexivity of this kind is 'reflex like'. It captures the unconscious and unintended consequences of industrial modernization, or what Beck labels the "self-confrontation" aspect of reflexive modernization (see Beck, 1994). In contrast, **second-order reflexivity** is about the self-critical and self-conscious reflection on processes of modernity, particularly instrumental rationality. It evokes a sense of agency, intention and change. Here actors reflect on and confront not only the self-induced problems of modernity, but also the approaches, structures and systems that reproduce them (Grin *et al.*, 2004; Stirling, 2006). It is this second-order reflexivity that is central to this paper.

Some scholars have taken agency a step further, by viewing second-order reflexivity as a mode of steering and coordination (Grin *et al.*, 2003; Grin, 2006; Meadowcroft, 1999). This is clearly what Voß & Kemp (2005, p. 8) have in mind, for example, when they define reflexive governance as: "the organisation (modulation) of recursive feedback relations between distributed steering activities". Steering for sustainability is conceptualized as a strategic process involving five key elements (Voß & Kemp, 2006, pp. 17–20): (i) transdisciplinary knowledge production; (ii) experiments and adaptivity strategies and institutions; (iii) anticipation of long-term systems effects of measures; (iv) interactive participatory goal formulation; and (v) interactive strategy development. As practical examples of such forms of steering, existing procedures are mentioned such as constructive technology assessment, deliberative policy making, social appraisal of technology, and Local Agenda 21 (Voß & Kemp, 2005; 2006). While such arrangements might represent potential sites of reflexivity, they seem to fall short of the kind of organized or modulated approach to steering so central to definitions of reflexive governance (for sustainability).

Specific conceptualizations of steering for sustainability are also are emerging from recent discussions on 'systems innovation' and 'transition management'. Scholars and practitioners—predominantly in the Netherlands—have been developing various frameworks to facilitate the transformation of large-scale socio-technological systems, for example, those associated with agriculture, transportation, health, communication and energy.[5] One such framework conceptualizes steering as a process of managing transitions towards more sustainable futures (Loorbach & Rotmans, 2006; Vollenbroek, 2002). A transition is understood as "... a gradual, continuous process of change where the structural character of a society (or complex sub-system of society) transforms" (Rotmans *et al.*, 2001, p. 16). The government's role is described as "plural" (Rotmans *et al.*, 2001, p. 26). On the one hand, state actors are called upon to steer systems innovation towards sustainable objectives (content role) whilst, on the other hand, they need to facilitate and evaluate procedures that mobilize and engage actors (process role).

For those interested in socio-technological innovations, steering for sustainability is conceptualized as emerging through partnerships and networks (Geels *et al.*, 2004; Kemp *et al.*, 1998). Under this account, steering involves creating platforms where different actors can work together *horizontally* to define visions and strategies. For example, Kemp *et al.* (1998) put forward the idea of *strategic niche management* to facilitate regime transformation. Niches are conceived as spaces where innovation is actively protected against certain aspects of the existing regime, such as unfavourable market conditions.[6] Niches are also zones where a variety of experiments take place, which facilitate learning and build networks (Kemp *et al.*, 1998, pp. 185–86).

This more bottom-up approach to steering more explicitly recognizes the role of interests and strategic action in technological development. The emphasis on actors and their networks also seems to reflect how collaboration works in observed practice (see Grin *et al.*, 2004, Hajer & Wagenaar, 2003). There are, of course, important differences between reflexive and network notions of governance to acknowledge here. In particular, actors in reflexive practices are doing much more than simply working together *around* and *across* existing modernist institutions (e.g. the state), as network governance scholars describe (e.g. Kooiman, 1993; Rhodes, 1997). In addition, they are engaging in a process of transforming the very systems in which they operate; a change they may or may not support. This added dimension of reflexive governance is likely to raise a unique set of political issues, some of which are explored below.

Practising Reflexive Governance in a Complex World

The discussion thus far has focused on particular interpretations of what steering for sustainability might entail. While these present useful starting points for appreciating reflexive governance, as conceptualizations of its practice they fall short in several respects. First, they tend to ignore or downplay the diverse and dynamic political landscape in which reflexive arrangements typically find themselves, as observed by several empirical studies (see Grin *et al.*, 2004; Loeber, 2004; Schot, 1998). Steering for sustainability is conceived largely as a more or less uncontroversial act, which is instigated by either the executive branch of government or innovators.[7] There is little sense of the 'political' dimensions of reflexive processes, and the 'politics' that they generate. Politics might be acknowledged, but it is implicitly seen as a bother to steering, rather than as something that governance needs to accommodate and build upon (cf. Lindblom, 1990, pp. 221–22; 226–30). Moreover given that *power* is distributed (a phenomenon reflexive governance scholars also emphasize), one crucial question that deserves more consideration is: how do attempts at steering—which seek to transform power relations—relate to the existing preferences and demands of elected officials, the legislature, the market and the public?

Secondly, existing conceptualizations of reflexive practice carry questionable assumptions about the way contemporary policy making works. For example, in transition management, the policy making is construed largely as a government-induced exercise that involves the willing engagement of relevant 'stakeholders' such as interest groups, scientists, businesses and experts.[8] However, this seems to underestimate the ambiguity over roles in contemporary politics, particularly the expectations and perceptions of the state.[9] It is not uncommon for state-based attempts to stimulate innovation for sustainability to meet resistance from non-state actors who fear co-option or loss of power (Hendriks, 2002; Sagoff, 1999; Thomas, 2003). Alternatively, existing or past controversies can affect the willingness of actors to participate in state-initiated projects (Grin *et al.*, 2004). Empirical research on reflexive arrangements also suggests that existing institutional configurations often hinder the capacity of actors to shift—as reflexivity demands—flexibly between roles (Grin *et al.*, 2004) and, in some contexts, state actors show a clear preference for substantive issues over procedural matters (e.g. Hendriks, 2006a).

This relates to our third concern that existing conceptualizations of reflexive governance do not attend to the complicated and intricate relationship between large-scale transitions and civil society. To a certain extent, civil society is instrumentalized; actors are expected to co-operate in reflexive governance and assist by reforming practices and structures. This, however, neglects an important dimension of reflexive action, where actors from civil society engage in sub-politics outside and against the state (Beck, 1992; Berejikian & Dryzek, 2000). It seems that in emphasizing a cooperative role for civil society, sustainability scholars might be undermining their own project. For it is often the radical voices in civil society that provide the necessary background conditions for wide-scale social transformation to occur (see Dryzek *et al.*, 2003).

These shortcomings are interpreted here as a lack of consideration of the socio-political contexts in which reflexive governance is embedded, as well as a lack of appreciation of the struggles involved in reconfiguring institutional arrangements. For the most part, reflexive practice is understood as something that can be stimulated by instigating a relatively uncontested programme or project. On the one hand this project-based conceptualization (which we ourselves also use) acknowledges that reflexivity often needs to be facilitated, fostered and 'niched' under existing socio-economic conditions. Yet, on the other hand, in focusing too heavily on 'the project', one risks losing sight of how these moments of reflexivity *influence* and are *influenced by* their broader context. Here, metaphors such as 'bounded' and 'niche' are misleading because they suggest that reflexive arrangements and their actors are somehow disconnected from the very context that they are trying to transform or retain. We ask whether this is possible and, indeed, desirable?

A Richer Conceptualization of Enacting Reflexive Governance

To bring the socio-political context into view, an analytical framework that situates reflexive projects in a broader democratic system is proposed (after Mansbridge, 1999). This system is conceptualized as a series of multiple interconnected and overlapping spheres of public discourse, or discursive spheres (see Hendriks, 2006b).[10] These are arenas (both in the literal and figurative sense) where discussion and debate issues occur in the political domain. Discursive spheres form when social and political pressures generate sites of communication where ideas and opinions are exchanged, debated and contested.[11] They include state and non-state venues such as parliaments, expert committees, stakeholder round-tables, community meetings, public seminars, the media and other sites of public conversation (Hendriks, 2006b).

The use of 'discourse' here needs further clarification. To be clear, the term is *not* being used in its Foucauldian sense to describe a shared set of constructs. Rather, discourse is used in a loosely Habermasian way to refer to a social communicative process where actors expose and discuss different viewpoints, ideas, stories and arguments. However, there is no intention to evoke the normative ideal of communicative action (as described in Habermas' discourse ethics). Instead, discourse is used here descriptively to capture a whole gamut of communication that extends well beyond rational argumentation and consensus to include story telling, rhetoric, agonism, contestation and dissent.[12]

A discursive sphere is therefore a venue where communication (in its broadest sense) occurs on a particular public problem. Most spheres have a predominant form of communication. Some are aimed at *formal* or structured forms of communication, such as scientific inquiry, arbitration or consensus. Other spheres promote a more *informal* exchange of ideas, for example via symbolic, rhetorical or antagonistic means. Then there are those discursive spheres that attempt to bring together a *mix* of actors accustomed to a wide range of discursive practices. Some spheres are also far more public and inclusive than others; some are initiated by the state, while others emerge from civil society. What they all share, however, is that they provide spaces for conversations on matters of public concern.[13]

Drawing on these ideas, reflexive arrangements are conceptualized as one discursive sphere surrounded by a series of overlapping arenas of public discourse, as depicted in Figure 1.

Conceptualizing reflexive practice in this way draws attention to interconnectivities. Sites of second-order reflexivity are situated and connected to existing spaces of political debate where first-order reflexivity unintentionally surfaces. In doing so, we are encouraged to explore the dialectic between different worlds of reflexivity. The framework proposed here also shifts the focus from 'the project' to the interfaces where different discursive spheres overlap. Interfaces can be explored through the actions and experiences of various actors who move in and out of the reflexive arrangements, such as government officials, interest representatives, scientists, activists and citizens. These encounters—or 'reflexivity interfaces'—provide insights into how actors straddle the demands of their existing institutional world, and the world they are trying to create. A study of 'reflexivity interfaces' can also inform broader debates on the meanings of democracy, legitimacy and accountability when politics is displaced to multiple locations (Held, 2000; Warren, 2002). In particular, the way multiple sites of politics interact, and how they might reinforce or dislocate conventional understandings and institutions of democracy.

To be sure, the neat overlapping nature of discursive spheres as depicted in Figure 1 is a theoretical interpretation of what is, in practice, a highly unstructured dynamic process. But as the article will now show, this conceptualization provides a useful framework for examining how a reflexive arrangement works within a

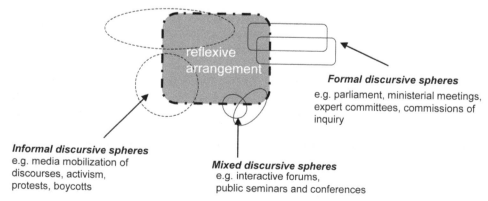

Figure 1. A reflexive arrangement in its discursive context

dynamic political landscape characterized by dispersed power and ambiguous roles.

The Gideon Project: Reforming Modernist Agricultural Practices

The case study relates to a particularly structured reflexive arrangement—the Gideon project (1995–96)—that was instigated by an advisory institution with close ties to the Dutch parliament. The Gideon project was an interactive evaluation of sustainable crop policies in the Netherlands, which contributed to a mid-term parliamentary review of a ten-year plan on agriculture reform (hereafter, the Plan).[14] Launched in 1991, the Plan was aimed principally at reducing the quantities of pesticides used per acre in the Netherlands. More broadly, it sought to reduce emissions to the environment and the Dutch agricultural system's "dependency" on chemical pesticides (LNV, 1990; Loeber, 2004). The addiction metaphor had been chosen deliberately, with the Plan explicitly emphasizing the fundamental nature of changes to agricultural practice it implied. As such, it represented one of the first policy steps targeted at radically transforming established Dutch agricultural practices (Bekke *et al.*, 1994; Wisserhof, 2000). Institutionally, the Plan had been produced outside the traditional framework of Dutch agriculture policy making. It also encompassed changes in the structures governing practices, most specifically legislation on pesticide admission. Simultaneously, there was a struggle over other matters, especially on whether to involve only traditional players or to extend participation to new players such as environmental movements, water managers and the Ministry of Environmental Affairs. In 1994, the Rathenau Institute, an independent adviser to parliament on scientific and technology issues, decided to contribute to this mid-term parliamentary review. It commissioned an external team to draft a proposal to evaluate the Plan on its behalf.[15] In its proposal, the team argued that in order to break through the 'deadlock' on pesticide dependency fundamental changes were needed not only in agricultural practices, but also in consumer behaviour, knowledge and technology development, agricultural policy making, and the social structures that governed these. To encourage this kind of second-order reflexivity the team proposed an 'interactive analysis'. The idea here was to facilitate a reconsideration of existing practices and structures by integrating discussion and analysis. Actors were encouraged to reflect collectively on the diagnosis of the deadlock and contribute strategies to resolving it.[16] In the end, the team's project design encompassed a series of preliminary studies, two rounds of interviews with relevant stakeholders and practitioners, as well as a future-orientation workshop, a work conference and an open day (see van Est *et al.*, 2002, pp. 117–20). Together, these interactive and reflexive aspects of Gideon (and the broader mid-term parliamentary review) make it a case of attempting reflexive governance in the context of distributed power and changing roles.

The Gideon Project in its Political Context

The Gideon project was surrounded by many venues of debate, negotiation and deliberation—some more public than others. In order to appreciate how it interfaced with this context, the paper first describes four key discursive spheres, as well as the Rathenau's role at the time as an 'intermediary' arena, intentionally established to operate at such interfaces. Figure 2 depicts this contextual landscape.

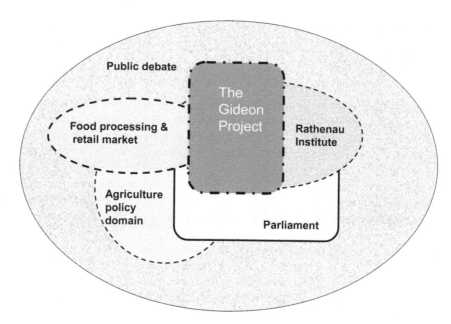

Figure 2. The Gideon project in its discursive context

Public debate.   Pesticides entered the public sphere as an issue of concern in the Netherlands during the late 1940s, as some agricultural experts began to raise questions about their potential health effects. However, it took Rachel Carson's (1962) *Silent Spring* to stimulate a broader wave of public concern about the risks of chemicals in the food chain and the environment. Pesticides were one of the key issues around which a broad societal movement on ecological concerns developed. By the 1970s the environmental movement, along with consumer groups, managed to bring other agriculturally related issues, such as production surpluses, manure emissions and animal welfare concerns, into the Dutch public sphere. Originally, the movement pressed for mitigating the effects of practices, which were otherwise left untouched. Yet, gradually a small organic farming sector emerged and pressure began to mount against modernist policies through consumer actions, environmental protests and lobbying (Priester, 2000). Public debate broadened even further with various environmental and food safety controversies; the outbreak of Bovine Spongiform Encephalopathy (BSE) in 2000 being a major stimulus.

The agriculture policy domain.   The Gideon project, as a reflexive arrangement, was contextualized by profound changes—both substantively and institution-ally—in the agricultural policy domain. The Netherlands' international reputation as a leader in agricultural production was due largely to the success of the 'iron triangle', which comprised: (i) the agriculture ministry, (ii) the *Landbouwschap* (a public–private agricultural organization with regulative authority), and (iii) agricultural specialists from Parliament (Bekke *et al.*, 1994; Wisserhof, 2000). The triangle was based on a strong consensus—partly maintained through personal commitments and unions of different kinds—on the modernization programme and the policies conceived to support it. 'Outsiders', like members of parliamentary delegations who were not agricultural specialists, were not allowed to have a say on these matters.

The discursive sphere on agricultural policy in the Netherlands began to open up when increasing criticism from the public sphere (see above) challenged the legitimacy of the modernization programme, particularly the influence of the iron triangle. When the Plan was launched in the early 1990s, the power had dispersed away from the iron triangle to a much wider set of governmental and non-governmental actors. When the Gideon project arrived on the scene in 1995 preparations were in place within government to terminate the existence of the triangle's central actor, the powerful *Landbouwschap*. In this era of profound change, ambiguity over roles prevailed.

Parliament.   Within the parliamentary discursive sphere, a monopoly of agricultural specialists had effectively depoliticized modernist agriculture practices for decades. However, from 1980 onwards this monopoly dissolved as a group of financial and environmental specialists entered the picture. Agriculture was then 're-politicized' as members of parliament from all sides of the political spectrum became involved in controversial debates, for example on agricultural subsidies, production restrictions or measures to reduce emissions or improve animal welfare. Here, too, fragmentation of power expressed itself and traditional roles began to destabilize. Agricultural specialists in Parliament lost their monopoly over agricultural policy, as actors with expertise on water and environmental management entered the picture. This led to a renegotiation of who could make decisions, when and on what basis (Bekke *et al.*, 1994). As Parliament increasingly became a discursive battle ground between modernists and reformers, its resulting policies showed both faces. For example, when the Plan was being discussed around 1991, Parliament adopted it virtually unchanged, despite vehement lobbying from traditional players. Simultaneously, it also passed a resolution demanding that government reach 'binding agreements' with agricultural interest organizations on the Plan's implementation.

The food processing and retailing market.   The food processing industry and retailers in the Netherlands initially ignored rising public concern about pesticides, and downplayed the significance of an emerging organic sector. The sector, however, was forced to become more discursive as public debate increased, and consumer preferences changed. Several retail groups, including the market leader, *Albert Heijn*, introduced 'ecological' (labelled) products to their assortment, and raised the environmental and animal welfare requirements on the food they sold. These changes were aimed primarily at maintaining or attaining a responsible public image. They were also about updating their market range to meet changing consumer preferences, which were perceived to be increasingly emphasizing quality, diversity and products' contribution to lifestyles and identities. Although market players were careful to maintain their autonomy, they sought contact with environmental, animal welfare and consumer organizations to ensure that their efforts satisfied critical groups. These efforts must also be viewed against a backdrop of subtly changing power balances. While the retailing sector gained more power over farmers and consumers through agricultural modernization (Priester, 2000), it began to realize that public concerns had the potential to disrupt existing production arrangements.

The Rathenau Institute as a mixed discursive sphere.   The Rathenau Institute, the instigator of the Gideon project, was founded in 1987 as an independent institute to provide advice to Parliament on science- and technology-related issues. Rathenau functions as a mixed discursive sphere to the extent that it brings together various political conversations from the public sphere, the parliament and the state. Although the Institute enjoys some degree of freedom, it nevertheless finds itself in the continuous process of legitimizing itself. When the Gideon project began the organization was undergoing a transformation of its mission from an analytical group providing policy advice on public concerns, to an organization contributing to societal debate. This shift in emphasis had two crucial implications. First, the quality of its projects would now be judged not only on the "quality of its reports *per se*, but also on its overall contribution to the arguments used in public debate and decision making" (van Eijndhoven, 2000, p. 155). Secondly, it became necessary to discuss methods to contribute to societal debates (Rathenau Institute, 1994), such as interactive technology assessment (see Grin *et al.*, 1997). Against this background, the Institute viewed the Gideon project both as a methodological testing ground, but also as a means to demonstrate its capacity to contribute to public debate.

Exploring Gideon at Interfaces

To appreciate how reflexive arrangements influence and are influenced by their dynamic political contexts, the discussion now moves to what happened as Gideon interfaced its surrounding discursive spheres, paying particular attention to processes of legitimization.

Interface with the market.   As the Gideon project connected with the market it met some resistance from a major retailing group. Facing the subtly changing power relations with consumers and civil society, the firm wanted to participate in the Gideon project to improve its public image. Initially, it engaged enthusiastically in the future-orientated workshop, but later rejected the jointly developed 'ecological' vision in the subsequent work conference (Loeber, 2004, pp. 235–36). This shift to a more defensive position can be explained in the context of earlier events. Prior to the Gideon project, the retailing group had organized a network of farmers to supply its retail operators with 'organic' produce. It adopted a new role as 'steward' and developed, in consultation with green groups, a particular version of sustainable agriculture. What resulted was a tight contracting system with certain primary producers, at the exclusion of many farmers. When the Gideon project arrived the retail group feared that supporting the ecological vision would further aggravate its relations with Dutch farmers. In the end, the firm's resistance, together with opposition from the *Landbouwschap*, meant that Gideon's ecological vision was abandoned.

Interfaces with the agricultural policy domain and the Rathenau Institute.   Against the backdrop of dispersing power and changing roles within and around the agricultural policy domain, Gideon posed challenges to existing notions of representation. In particular, the *Landbouwschap* was concerned that farmers included in the Gideon project were not 'representative' of the agricultural sector—a claim the *Landbouwschap* themselves made. A key methodological element of the Gideon project involved engaging people from the 'shop floor' in the project.[17] Thus, farmers were invited to participate as opposed to formal representatives of farmers' organizations, operational water managers rather than their directors and so on. This selection procedure was aimed at promoting creativity and encouraging alternative forms of knowledge. It was also intended to protect the project from many of the strategic games prevalent around the Plan process, and in the agricultural policy arena more generally.

Strategic players were given a place in a 'second ring'—an Advisory Committee—where they were invited to provide input to the Rathenau Institute on the project's proceedings. Many interest representatives willingly took up this role because it presented an opportunity through which they could influence the project. One of the primary issues on which these actors focused was the selection of participants (Loeber, 2004). On several occasions, various advisory committee members put forward strong suggestions for particular individuals to participate, for example, strategic players, a selection of 'shop-floor' representatives who actually held responsibilities in their constituencies, or their favoured experts. Some members even suggested that the Rathenau Institute was paying too little attention to accountability *vis-à-vis* 'the' sector by arguing that there is "a danger in giving the impression that one talks about the agricultural sector without consulting the sector itself".[18]

Of course, part of the reason that such pressure could be exerted at these moments was the Rathenau Institute's mission and its position *vis-à-vis* the agricultural sector. To deal with such interventions, careful manoeuvring was required by the project team, the Institute's project leader and the work conference facilitator (Loeber, 2004, pp. 226–37). For example, they reminded the participants and the Board members of their commitment to deliberative norms, such as accountability and reciprocity, and pointed out that the project's impact would be maximized if it worked with the problem definition set by its addressee, the Parliament. They also frequently drew attention to external developments that supported reform, such as changing consuming preferences, the recognition of these trends by powerful market parties and changing EU policies. At the work conference, such trends were brought prominently to the fore through plenary talks at the beginning of the two-day session. These exchanges helped to create results that were both reasonably innovative and met some acceptance, especially by remnants of the iron triangle.

These gains, however, did not prevent traditional players from trying to delegitimize the final report. During the Open Day, some participants claimed that the report contained little news. In the advisory board, some members deemed the report, with its rather "qualitative" approach, rather vague. The representative of the *Landbouwschap*, for instance, felt that it was sort of a "Christmas message put forward by the queen: all good intentions and best wishes, without practical ideas and means to show for it" (Loeber, 2004, p. 237). Other members, favouring radical reform, believed that the report would have gained more credibility had it included more quantitative analysis. They used this opportunity to promote particular experts and approaches, trying to strengthen their favoured position in an ongoing debate within the agricultural knowledge infrastructure.

Gideon's interface with the agricultural policy domain also worked the other way around. The project managed to have an impact on emerging policies, especially in its final stages and after-life. For example, various actors within the ministry have used the Gideon project to support more radical policy proposals, and to win over more conservative colleagues. As one policy maker explained: it was now clear that "they were not the only ones who think this approach is necessary and feasible" (Loeber, 2004, p. 245). Similarly, an information centre used Gideon to legitimate the options it had proposed for a follow-up policy to the Plan (IKC, 1998). The project leader of this plan justified this action saying that the Gideon project had made visible support for fundamental changes in crop protection (Loeber, 2004, p. 248). Finally, a water manager referred to Gideon to legitimize a more central role for water management in spatial planning (Faassen, 1999).

Interface with Parliament.   The Gideon project was intended to contribute to the mid-term Parliamentary review of the Plan. Although this fact was used frequently to deal with the strategic action of powerful actors, in the end the parliament paid little explicit attention to the final report. It simply accepted the minister's recommendations (which were based heavily on the Gideon report) without any parliamentary debate on what they entailed. On closer examination the parliament's use of Gideon was selective and opportunistic. On the one hand the project was cited positively to legitimize the adoption of support measures for crop protection in smaller cultivations—an issue some parliamentarians had been pushing for a while. On the other hand, Gideon was criticized by certain parliamentary actors who were keen to maintain good relations with powerful agricultural players. For example, at the Open Day, an agricultural specialist from parliament with strong ties to the traditional iron triangle raised again the accountability issue by casting doubt on whether the project reflected 'the' view from 'the' primary sector. Here, one sees clearly the dual-positioning of the Dutch parliament towards reflexive governance: it recognized the need to support long-term and radical objectives for sustainability, but at the same time it was bound historically and discursively to the demands of traditional actors.

Insights from Gideon for Reflexive Governance

For some, the Gideon project might be interpreted as just another inquiry or information-gathering exercise for Parliament. However, we prefer to interpret it as an attempt at reflexive governance—a reflexive moment amid a sea of everyday politics. Gideon's reflexive elements pertain to its attempt to reconsider the practices and structures of Dutch agriculture.

This study of Gideon highlights the kinds of political struggles that reflexive arrangements can generate, especially when situated in close proximity to the state. The case also illustrates the strong interconnectivities between reflexive arrangements and their surrounding discursive spheres. Most 'discursive activity' occurred as Gideon interfaced with the agricultural policy domain and the market. The analysis reveals that at these interfaces a dialectic occurred: the Gideon project both *influenced*, and was *influenced by*, its surrounding context. On the one hand, the project had an impact—albeit indirect—on the agricultural policy domain. On the other hand, the project was shaped and influenced by surrounding discursive activities particularly those of traditional, powerful players. Interestingly, this dialectic resulted as actors attempted to legitimize their new positions and policies in a situation of dispersing power and changing roles. Both the retailing group and the *Landbouwschap*'s influence on the project were induced by their anticipations of how the project might affect the legitimacy of their evolving positions. Similarly, actors from the agricultural and water domain used Gideon to legitimize the policy innovations they proposed.

Another related insight from Gideon is the 'politics' that reflexive arrangements incite. Various political tensions surfaced as actors and institutions tried to juggle the competing demands of second-order reflexivity and their strategic interests in other discursive spheres. For example, both the parliament and the Rathenau Institute positioned themselves in a dual manner: they were committed to profound substantive change in the agriculture sector, but they also remained accountable to the demands of traditional 'iron triangle' players. Politics was not only present in the Gideon project, but it was also accommodated. The Rathenau Institute and the Gideon team worked hard to accommodate changing power relations and roles, for example by addressing various methodological and substantive concerns raised by different actors. They also tried to anticipate potential objections and, wherever possible, they adapted the project's design and content to resolve external concerns. This suggests that reflexive arrangements—whether structured or unstructured—need to be co-ordinated by institutions with the capacity to balance and accommodate the politics that second-order reflexivity is likely to generate (cf. Roep *et al.*, 2003).

The final insight to highlight from this case study is that the impact of reflexive arrangements can be subtle and indirect. They rarely culminate in grand policy decisions made by formal institutions, such as relevant ministries or even the Parliament. As the Gideon project shows, formal discursive venues such as Parliament tend to play an inactive role in the everyday activities of reflexive policy making, especially when radical reforms are involved. In this case parliament may have been inactive, but it was certainly not insignificant. It provided the locus around which much discursive activity was centred, and this helped to legitimate the project. While, ultimately, the Gideon project was eventually able to lend sufficient legitimacy to visions and strategies for fundamental reform both in the agricultural sector and beyond, this analysis shows that a lot of 'hard work' and politics was involved in reaching that point.

Conclusions

Steering for sustainability can be understood as reflexive governance—a process of fundamentally reconsidering the way our socio-technical systems are structured, practised and most significantly governed. How this reflexive mode of steering works alongside the dynamic world of contemporary politics has been the central question of this paper. To study this empirically we have elaborated on a conceptualisation of steering that situates sites of reflexivity (reflexive arrangements) within their broader socio-political context, including multiple arenas, actors and forms of political communication. What is novel about this conceptualization is that it connects emerging literature on steering for sustainability with relevant ideas in democratic theory. The product of this fruitful exchange is a heuristic device for exploring how societies can be steered towards sustainability within the context of a dynamic governance landscape.

The analysis of the Dutch Gideon project demonstrates how intentional sites of reflexivity can *influence* and in some cases *challenge* other arenas of political debate, and vice versa. While the precise ways in which this occurs may vary over contexts and political systems, any attempt to facilitate reflexive governance will involve some form of political struggle. Tensions arise as actors try to reconcile the demands of reflexivity (being open, self-critical and creative) with the demands of their existing political world (closed preferences, agenda driven, control). Depending on the strategic interests of these actors, attempts at reflexive governance will be championed or resisted (see Hendriks, 2006a). In the Gideon project one can see how some actors threatened by second-order reflexivity sought to undermine the process by calling into question the accountability of reflexive practitioners, and the legitimacy of their proposals. In contrast, strategic actors who saw value in Gideon used the process and its outcomes to promote and legitimize their own positions and actions.

How does this study of a one-off reflexive arrangement speak to those interested in sustained long-term steering programmes for sustainability? One key message to take from this analysis is that any attempt to steer societies towards sustainability will involve and affect politics. This study also reminds us of a broader phenomenon in contemporary governance: that democracy increasingly attains its meaning in multiples locations (Latour, 2005; Mol, 2002). Democratic concepts, such as accountability and legitimacy, are not 'givens' in the context of reflexive governance. Instead, they are constructed and used strategically as resources by actors, for example, who call into question 'the accountability' of reflexive practitioners, or 'the legitimacy' of their recommendations. While, ultimately, the Gideon project was able to lend sufficient legitimacy to visions and strategies for fundamental reform both in the agricultural sector and beyond, the analysis shows that a lot of 'hard work' and politics was involved in reaching that point. Given that a one-off arrangement like Gideon involved so much resistance, tension and political manoeuvring, there is much more work to do to before the political and democratic implications of larger more radical programmes for sustainability reform can be fully appreciated.

Acknowledgements

This article is based on research conducted within the Dutch knowledge Network on System Innovations (KSI; www.ksinetwork.nl <http://www.ksinetwork.nl/>). For their comments, we would like to thank two anonymous reviewers, Nicholas Buchanan, Jochen Monstadt, Jens Newig, Jan-Peter Voß, Lydia Sterrenberg and participants of the Governance for Suatainable Development Workshop in Berlin, February 2006 (organized by the Sozial-ökologische Forschung).

Notes

1. Our consideration of ambiguity is broader than that of other scholars in this contribution. In addition to ambiguity over knowledge claims and definitions (for example of 'sustainability'), we are also referring to ambiguity over the roles of actors, institutions and structures.
2. For details on the distinction between weak and strong forms of ecological modernization, see Christoff (1996).
3. We support the idea that reflexive governance for sustainability requires a broader over-arching strategic process (see Voß & Kemp, 2006), but believe that in practice these are rare. The starting point for our discussion is thus empirical rather than normative.
4. In one comprehensive inventory, Lynch (2000) identified six broad categories of reflexivity: mechanical, substantive, methodological, meta-theoretical, interpretative and ethnomethodological. In relation to sustainability, reflexivity is applied mostly in its substantive sense (as a social phenomenon of late modernity) and in a methodological sense (as a cognitive process of reflection/self-reflection).

5. See www.ksinetwork.nl.

6. For our purposes here, a regime is best understood as the dominant socio-technological structure.

7. For example, Rotmans *et al.* (2001, p. 25) described how the state should assume "a leading role . . . [n]ot by acting as the great commander, enforcing change, but by inspiring a collective learning process and encouraging others to think along and participate". Others stress how the state should modulate ongoing developments (Kemp & Rotmans, 2004) or create transition partnerships with autonomous and creative thinkers (Loorbach & Rotmans, 2006). Other accounts of 'steering' towards sustainability, particularly those interested in the development of innovative technologies, emphasize partnerships and collaboration amongst actors. Authors here explicitly acknowledge that steering: ". . . is not just something for governments: industry and NGOs are well placed to initiate and run niche projects" (Kemp *et al.*, 1998, p. 189).

8. In their discussion of transition management, Rotmans *et al.* (2001, p. 30) asserted that: "All social actors look to the government to take the lead".

9. See note 1.

10. In contrast, Jane Mansbridge (1999) conceptualized the deliberative system as a *spectrum*. This idea has been extended into a system of interconnected discursive spheres, for reasons elaborated on elsewhere (see Hendriks, 2006b).

11. These pressures might be generated by one-off crises (e.g. the BSE crisis) or they might emerge slowly over time (e.g. climate change).

12. This understanding of discourse is perhaps closer to Habermas' earlier sociological diagnosis of the public sphere in late capitalism, what Benhabib (1992, pp. 84–95) labelled as his "discourse model of public space" in contradistinction to his later work on the "moral theory of discourse ethics". Our usage of discourse also overlaps with certain interpretations of Hannah Arendt's concept of public discourse, which highlight *expressive* and *communicative* modes of action (Benhabib, 1992, pp. 74–81; 2003, pp. 123–30; d'Entréves, 1994, pp. 84–5).

13. Here the concept of the 'discursive sphere' is broader than the notion of 'public sphere' which is used commonly in political theory to refer to those arenas where issues are brought into the public domain, that is, 'made public' (Benhabib, 1992; Fraser, 1992). Discursive spheres encapsulate much more than this by including both public and non-public venues of discussion.

14. Although responsibility for the following discussion is fully ours, we are indebted for discussion of this project by Anne Loeber, who analysed documents internal and external to the project, and conducted interviews with members of the project team, the Rathenau Institute, and stakeholders involved in the project (see Loeber, 2004).

15. The team comprised two university groups (a STS and a public policy studies group) from outside the agricultural domain, as well as a young, independent institute at the fringes of that domain, CLM (the Centre for Agriculture and Environment). One of the authors of this paper (Grin) was a project leader of this team.

16. This approach (see Grin & van de Graaf, 1996) was inspired by Guba & Lincoln's (1989) method of constructivist evaluation.

17. Additional selection criteria included that participants should: represent a variety of positions on the issue; stem from various parts of the economic chain; stem from different agricultural sectors; and be prepared to accept deliberative norms.

18. Minutes from an advisory board meeting, 12 December, 1995. Internal document, Rathenau Institute.

References

Alexander, J. C. (1995) *Fin de Siècle Social Theory. Relativism, Reduction and the Problem of Reason* (London & New York: Verso).

Arts, B. & van Tatenhove, J. (2004) Policy and power: a conceptual framework between the old and new policy idioms, *Policy Sciences*, 37, pp. 337–356.

Beck, U. (1992) *Risk Society: Towards a New Modernity* (London: Sage).

Beck, U. (1994) The reinvention of politics: towards a theory or reflexive modernization, in: U. Beck, A. Giddens & S. Lash (Eds) *Reflexive modernization: politics, tradition and aesthetics in the modern social order*, pp. 1–55 (Cambridge: Polity Press).

Beck, U., Bonss, W. & Lau, C. (2003) The theory of reflexive modernization: Problematic, hypotheses and research programme, *Theory, Culture and Society*, 20(1) pp. 1–33.

Bekke, H., de Vries, J. & Neelen, G. (1994) *De salto mortale van het ministerie van Landbouw, Natuurbeheer en Visserij. Beleid, organisatie en management op een breukvlak* (Alphen aan den Rijn: Samson H.D. Tjeenk Willink).

Benhabib, S. (1992) Models of public space: Hannah Arendt, the liberal tradition, and Jürgen Habermas, in: C. Calhoun (Ed.) *Habermas and the public sphere*, pp. 73–98 (Cambridge MA: MIT Press).

Benhabib, S. (2003) *The reluctant modernism of Hannah Arendt* (Lanham: Rowman & Littlefield Publishers Inc).

Berejikian, J. & Dryzek, J. S. (2000) Reflexive action in international politics, *British Journal of Political Science*, 30, pp. 193–216.

Carson, R. (1962) *Silent Spring* (Boston: Houghton Mifflin).

Christoff, P. (1996) Ecological modernisation, ecological modernities, *Environmental Politics*, 5(3), pp. 476–500.

Considine, M. (2000) Contract regimes and reflexive governance: Comparing service reforms in the United Kingdom, the Netherlands, New Zealand and Australia, *Public Administration*, 78(3), pp. 613–638.

Cunliffe, A. L. & Jun, J. S. (2005) The Need for Reflexivity in Public Administration, *Administration in Society*, 37(2), pp. 225–242.

'Entréves, M. P. (1994) *The political philosophy of Hannah Arendt* (London: Routledge).

Dryzek, J. S., Downes, D., Hunold, C., Schlosberg, D., with Hernes, H.-K. (2003) *Green states and social movements: Environmentalism in United States, United Kingdom, Germany and Norway* (Oxford: Oxford University Press).

Faassen, R. (1999) Landbouwkundige ontwikkelingen als sleutel tot integraal ruimtegebruik, in: Raad voor het Milieu- en Natuuronderzoek (RMNO) *Ingrediënten voor een duurzame samenleving. Twaalf essays over integraal ruimtegebruik*, pp. 85–93 (Alphen aan den Rijn: Samson H.D. Tjeenk Willink.)

Fraser, N. (1992) Rethinking the public sphere: A contribution to the critique of actually existing democracy, in C. Calhoun (Ed.) *Habermas and the public sphere*, pp. 109–142 (Cambridge: MIT Press).

Geels, F. W., Elzen, B. & Green, K. (2004) General introduction: Systems innovation and transitions to sustainability, in B. Elzen, F. W. Geels & K. Green (Eds) *Systems innovation and the transition to sustainability: Theory, evidence and policy*, pp. 1–16 (Cheltenham: Edward Elgar).

Grin, J. (2006) Reflexive modernization as a governance issue: or designing and shaping re-structuration. in: J.-P. Voß, D. Bauknecht & R. Kemp (Eds) *Reflexive governance for sustainable development*, pp.57–81 (Cheltenham: Edward Elgar).

Grin, J., Felix, F., Bos, B. & Spoelstra, S. (2004) Practices for reflexive design: Lessons from a dutch programme on sustainable agriculture, *International Journal of Foresight and Innovation Policy*, 1(1–2), pp. 126–149.

Grin, J., van de Graaf, H. & Vergragt, P. J. (2003) Een derde generatie milieubeleid: een sociologisch perspectief en een beleidswetenschappelijk programma, *Beleidswetenschap*, 17(1), pp. 57–71.

Grin, J. & van de Graaf, H. (1996) Technology Assessment as learning, *Science, Technology and Human Values*, 20(1), pp. 72–99.

Grin, J., van de Graaf, H. & Hoppe, R. (1997) *Technology assessment through interaction: A guide* (The Hague: The Rathenau Institute).

Guba, E. & Lincoln, Y. (1989) *Fourth Generation Evaluation* (Seven Oaks: Sage).

Hajer, M. (2003) Policy without polity? Policy analysis and the institutional void, *Policy Sciences*, 36(2), pp. 175–195.

Hajer, M. & Wagenaar, H. (Eds) (2003) *Deliberative policy analysis: Understanding governance in the network society* (Cambridge: Cambridge University Press).

Held, D. (2000) The changing contour of political community: rethinking democracy in the context of globalization, in: B. Holden (Ed.) *Global Democracy: Key Debates*, pp. 17–31 (London: Routledge).

Hendriks, C. M. (2002) Institutions of deliberative democratic processes and interest groups: Roles, tensions and incentives, *Australian Journal of Public Administration*, 61(1), pp. 64–75.

Hendriks, C. M. (2006a) When the forum meets interest politics: strategic uses of public deliberation, *Politics and Society*, 34(4), pp. 1–32.

Hendriks, C. M. (2006b) Deliberative integration: Reconciling civil society's dual role in deliberative democracy, *Political Studies*, 5(3), pp. 486–508.

IKC (1998) *Gewasbescherming in een nieuw tijdperk. Verkenningen van toekomstige beleidsopties Gewasbescherming* (The Hague: Ministry of Agriculture.)

Kemp, R. & Rotmans, J. (2004) Managing the transitions to sustainable mobility, in: B. Elzen, F. W. Geels & K. Green (Eds) *Systems Innovation and the Transition to Sustainability: Theory, Evidence and policy*, pp. 137–167 (Cheltenham: Edward Elgar).

Kemp, R., Schot, J. & Hoogma, R. (1998) Regime shifts to sustainability through processes of niche formation: The approach of strategic niche management, *Technology Analysis & Strategic Management*, 10(2), pp. 175–195.

Kooiman, J. (Ed.) (1993) *Modern governance: new government–society interactions* (London: Sage).

Latour, B. (2005) From Realpolitik to Dingpolitik, in: B. Latour & P. Weibel (Eds) *Making Things Public— Atmospheres of Democracy*, pp. 4–31 (Karlsruhe & Cambridge: ZKM & MIT Press).

Lindblom, C. E. (1990) *Inquiry and Change. The troubled attempt to understand and shape society* (New Haven: Yale University Press).

LNV (1990) *Meerjarenplan Gewasbescherming* (The Hague: Ministry of Agriculture, Nature Conservation and Fisheries).

Loeber, A. (2004) Practical wisdom in the risk society, PhD thesis, The Political Science Department, The University of Amsterdam.

Loorbach, D. & Rotmans, J. (2006) Managing Transitions for Sustainable Development, in: X. Olsthoorn & A. Wieczorek (Eds) *Understanding Industrial Transformation Views from different disciplines*, pp. 187–206 (Leusden: Springer).

Lynch, M. (2000) Against reflexivity as an academic virtue and source of privileged knowledge, *Theory, Culture & Society*, 17(3), pp. 26–54.

Mansbridge, J. (1999) Everyday talk in the deliberative system, in: S. Macedo (Ed.) *Deliberative politics— essays on democracy and disagreement*, pp. 211–239 (Oxford: Oxford University Press).

Meadowcroft, J. (1999) Planning for sustainable development: what can we learn from the critics?, in: M. Kenny & J. Meadowcroft (Eds) *Planning Sustainability*, pp. 12–38 (London: Routledge).

Mol A. (2002) *The Body Multiple: Ontology in Medical Practice* (Durham: Duke University Press).

Pierre, J. (Ed.) (2001) *Debating governance: Authority, steering and democracy* (Oxford: Oxford University Press).

Priester, P. R. (2000) Landbouw—Part 1b, in: J. Bieleman (Ed.) *Techniek in Nederland in de twintigste eeuw: Landbouw & Voeding*, pp. 65–125 (Zutphen: Walburg Pers).

Rathenau Institute (1994) *Annual Report 1994: The Rathenau Institute and debate* (The Hague: The Rathenau Institute).

Rhodes, R. A. W. (1997) *Understanding governance: Policy networks, governance, reflexivity, and accountability* (Buckingham: Open University Press).

Roep, D., van der Ploeg, J. D. & Wiskerke, J. S. C. (2003) Managing technical–institutional design processes: some strategic lessons from environmental co-operatives in the Netherlands, *Netherlands Journal of Agrarian Studies*, 51(1–2), pp. 195–217.

Rotmans, J., Kemp, R. & van Asselt, M. (2001) More evolution than revolution: Transition management in public policy, *Foresight*, 3(1), pp. 15–31.

Sagoff, M. (1999) The view from the Quincy library: Civic engagement and environmental problem solving, in: R. K. Fullinwider (Ed.) *Civil society, democracy and civic renewal*, pp. 161–183 (New York: Rowan Littlefield).

Schot, J. (1998) The usefulness of evolutionary models for explaining innovation. The case of the Netherlands in the nineteenth century, *History and Technology*, 14, pp. 173–200.

Stirling, A. (2006) Precaution, foresight and sustainability, in: J.-P. Voß, D. Bauknecht & R. Kemp (Eds) *Reflexive governance for sustainable development*, pp. 225–272 (Cheltenham: Edward Elgar).

Thomas, C. W. (2003) Habitat conservation planning, in: A. Fung & E. O. Wright (Eds) *Deepening democracy: Institutional innovation in empowered participatory governance*, pp. 144–172 (London: Verso).

van Eijndhoven, J. (2000) The Netherlands: Technology Assessment from Academically Oriented Analysis to Support of Public Debate, in: N. J. Vig & H. Paschen (Eds) *Parliaments and Technology. The Development of Technology Assessment in Europe*, pp. 147–172 (Albany, New York: SUNY Press).

van Est, R., van Eijndhoven, J., Arts, W. & Loeber, A. (2002) The Netherlands: Seeking to Involve Wider Publics in Technology Assessment, in: S. Joss & S. Bellucci (Eds) *Participatory Technology Assessment: European Perspectives*, pp. 108–125 (London: Centre for the Study of Democracy).

Vollenbroek, F. (2002) Sustainable development and the challenge of innovation, *Journal of Cleaner Production*, 10, pp. 215–223.

Voß, J.-P., Bauknecht, D. & Kemp, R. (2006) (Eds) *Reflexive governance for sustainable development* (Cheltenham: Edward Elgar).

Voß, J.-P. & Kemp, R. (2005) Sustainability and reflexive governance: Incorporating feedback into social problem solving. Paper presented at International Human Dimensions Programme on Global Environmental Change (IHDP) Open Meeting, Bonn, 9–13 October.

Voß, J.-P. & Kemp, R. (2006) Sustainability and reflexive governance: Introduction, in: J.-P. Voß, D. Bauknecht & R. Kemp (Eds) *Reflexive governance for sustainable development*, pp. 3–28 (Cheltenham: Edward Elgar).

Warren, M. E. (2002) What can democratic participation mean today?, *Political Theory*, 30(5), pp. 677–701.

Wisserhof, J. (2000) Agricultural policy making in the Netherlands: Beyond corporatist policy arrangements? in: J. van Tatenhove, B. Arts & P. Leroy (Eds) *Political modernisation and the environment. The renewal of environmental policy arrangements*, pp. 175–198 (Dordecht: Kluwer Academic Publishers).

Moving Outside or Inside? Objectification and Reflexivity in the Governance of Socio-Technical Systems

ADRIAN SMITH & ANDY STIRLING

Introduction

The case for considering the co-evolution of social and technological systems in sustainable development is well made in the literature on Science and Technology Studies and Science, Technology and Society (collectively abbreviated as STS; for reviews see Rip & Kemp (1998) or Russell & Williams (2002)). A variety of studies have shown how social practices and technological artefacts shape and are shaped by one another (e.g. Geels, 2002; Latour, 1996; Raven, 2006; Shove, 2003). The ways these apparently seamless 'socio-technical' developments unfold suggest sustainability governance must bring its 'technology-fix' and 'behavioural-change' tendencies into better correspondence (Brand, 2005).

As a consequence of these co-evolutionary insights, a number of scholars recommend governance practitioners adopt a 'socio-technical systems' perspective in order to guide development in more sustainable directions (Elzen *et al.*, 2005; Kemp *et al.*, 1998). However, the relationship between governance processes and socio-technical systems has received little systematic attention. Whilst analyses of socio-technical systems recognize that their 'systems' remain open to considerable social interpretation and construction, there nevertheless exists a tendency to objectify and fix these systems when recommending governance interventions (Rotmans & Kemp, 2001). Confusion prevails between reflexive interpretations of socio-technical systems and objectifying injunctions for authoritative interventions (see also Latour, 2004).

Following arguments for reflexivity put by others (Hajer & Wagenaar, 2003; Voβ *et al.*, 2006), this paper presents a novel scheme for thinking systematically about governance in relation to socio-technical systems. Our heuristic scheme is built around two ideal-typical representations of governance in relation to the socio-technical. The first ideal-type, called 'governance on the outside', considers governance subjects attaining an objective distance from a discrete, uniquely knowable socio-technical system. The second ideal-type, called 'governance on the inside', is more reflexive and recognizes how framings of the system by different actors, and their inter-subjective negotiation in governance arenas, effectively involves those arenas in the (social) (re)construction of the socio-technical 'system' itself. Rather than governance being outside the system (our first ideal-type), the governance arena and the socio-technical system are internal to one another.

Governance is conceptualized somewhat differently under our scheme compared to the mainstream literature (e.g. Kooiman, 2003; Rhodes, 1997; Stoker, 1998). Processes of social appraisal and social commitment are brought to the fore in the analysis (second section)—recognizing that (like more conventional accounts of governance) these operate through networks of actors and institutions, and that these processes determine authoritative attempts to bring 'socio-technical system' developments into line with public goals. Highlighting appraisal and commitment uncovers important sources of tension between objectifying and reflexive modes of governing socio-technical systems. The two ideal-typical modes of governance enable a systematic exploration of these tensions. 'Governance on the outside' is managerial in its strategies and is consequently susceptible to governance challenges around insufficiently authoritative appraisal (third section). 'Governance on the inside' is more political in approach, but must confront the considerable challenges of doing reflexive governance (fourth section). Each lends itself to different yet coherent strategies for addressing sustainability issues central to this journal special issue, of ambivalence, uncertainty and distributed power (fifth section). The paper is not simply re-stating the top-down/bottom-up debate in the environmental planning literature. Rather, it seeks to provide a novel way for analysts (and practitioners) to appreciate and live with the contradictory tendencies in governance to simultaneously objectify that which it governs whilst permitting (or struggling to exclude) a diversity of framings of the governed.

In practice, governance is rarely characterized purely as one of the two ideal types set up here ('outside' or 'inside'). Many accounts of governance wrestle with the tensions identified but remain unclear as to their sources. The analysis identifies tensions as arising from the interplay of contrasting strategies associated with more managerially inclined governance attempts to objectify socio-technical systems, and those associated with more politically inclined governance that tries reflexively to open up and negotiate various 'system' framings held by different groups (sixth section). Although objectification can facilitate co-ordinated interventions and management, this is only robust *where actors agree over the object around which they are supposed to be co-ordinating*. Instead, a research agenda is identified that explores governance dynamically as conflicting, strategic attempts towards objectification and reflexivity in steering socio-technical systems. In so doing, a heuristic is proposed for exploring a wide and diffuse literature and for thinking about reflexive governance more rigorously.

The Governance Challenge: Appraising Socio-Technical Systems and Committing to Change

A socio-technical systems perspective has come to the fore in studies of technology and sustainability (Elzen *et al.*, 2005; Weber & Hemmelskamp, 2005). The governance challenge is no longer simply to promote cleaner technologies. Instead, it lies in transforming wider socio-technical systems (Berkhout, 2002). The new focus recognizes that technologies are embedded within broader socio-political and economic networks. Some of the reasons cleaner technologies are not diffusing more rapidly relate to overarching structures of design criteria and routines, markets, final consumer demand, institutional and regulatory systems, and inadequate infrastructures for change. Technology developers have limited room for unilateral manoeuvre in relation to these system-level factors. Reinforcing this focus at the broader socio-technical level is a realization that radical changes at a whole system scale are needed to deliver the revolutionary material efficiencies and emission reductions that sustainability demands (Rotmans & Kemp, 2001).

Since successful socio-technical development emerges through complex networks of actors, artefacts and institutions, so governance will need to engage across many of the points and processes within those networks (Smith *et al.*, 2005). Imposing normative goals of sustainability upon existing systems implies connecting and synchronizing changes among a formidable array of processes across many different points in the system. Governance must consequently fulfil distinct diagnostic, prognostic, prescriptive and co-ordination functions. First, it must identify problems of unsustainability in the socio-technical system. Secondly, it must look forward through the complex and uncertain dynamics of interpenetrating and tightly-coupled technological, social and environmental systems. Thirdly, it must develop a set of shared normative criteria with which to appraise the best governance responses to these problems. Fourthly, it must implement these solutions by forming commitments, and intervening to effect change. Fulfilment of these functions is far from straightforward. A wide variety of distributed knowledges, skills and other resources (e.g. technical, finance, legitimacy, authority) must be marshalled if socio-technical development is coherently to be comprehended and steered. In exercising these functions, governance must be highly adaptable, since

> the dynamics of many socio-technical processes are such that the matching governance practices seem to be continuously 'out of breath': they have been overtaken by the developments, because the developments are more dynamic and the governing is not dynamic (enough) (Kooiman, 1993, p. 36).

Amongst these developments are the unintended consequences of earlier governance interventions (Voβ & Kemp, 2006).

However governance is characterized, it necessarily involves two complementary, intertwined and mutually co-constituting (but analytically distinct) processes: "social appraisal" and "social commitment" (Stirling, 2005; 2008).

Since our ideal-types for governing socio-technical systems are based around different understandings of the relationship between the socio-technical, its social appraisal, and social commitment (third and fourth sections), they merit consideration here.

Social appraisal comprises essentially 'epistemic' processes—'ways of understanding' the socio-technical system. Here, knowledge is constructed, imbued with meaning and subject to social learning (Nowotny *et al.*, 2001; Webler *et al*, 1995; Wynne, 1995). Social commitments, by contrast, involve more 'ontological' 'ways of being' in relation to the socio-technical system (Feenberg, 2002; Leach *et al.*, 2005; Wynne, 2001). Here, real relationships are formed, tangible resources produced and deployed and concrete governance interventions undertaken. Seen in this way, the recursive interlinkage of appraisal and commitment cross-cuts other, more elaborate, taxonomies through which to understand governance functions in the steering of socio-technical systems. For instance, activities of 'problem identification', 'goal formulation', and 'strategy implementation' (Voβ & Kemp, 2006) may each be seen to comprise elements both of appraisal and commitment in these senses. By highlighting this straightforward distinction between appraisal and commitment, the hope is to focus specifically on the important but neglected contrast between managerial and political perspectives.

Commitments

Social actors can be 'committed' to a socio-technical system in a number of overlapping and related senses. One form of commitment is rooted in material interests, in the sense that actors rely on the system to satisfy some need, e.g. housing, food, clothing, mobility, entertainment, lighting, warmth. A more direct and constitutive form of material commitment is the way that the functioning of the system itself requires the co-ordinated mobilization of actors and resources. Obvious examples include capital finance, operational subsidies, infrastructure rights, contractual security, regulatory protection, political patronage, support in knowledge production, intellectual property guarantees, user demand, promotion of learning, facilitation of recruitment or measures for the externalization of liability. The way in which actors engage in the production and reproduction of these relationships and resources represents a distinct form of commitment to particular configurations of the socio-technical system in question.

Obviously, the commitments above relate closely to the reciprocal benefits deriving from participation in the reproduction of the system. Such benefits include the generation of a financial profit or furthering some other institutional interest. Consumers in the socio-technical system derive benefits from the efficiency, convenience or style with which the system helps them meet their needs, at the same time as furnishing a demand (and resources) that shapes reproduction of the system (Shove, 2003). Governments commit to the system because it facilitates further economic development, fosters social cohesion, or contributes to policy goals or displays more specific, expedient political benefits, whilst state regulation and enforcement of property rights contribute to system reproduction. Citizens commit to socio-technical systems because they facilitate civic and community activity over and above individual consumption benefits, whilst conferring a degree of legitimacy for system reproduction. However, commitments need not be rigid; if an alternative socio-technical configuration provides equivalent benefits in some preferable way, then commitments may shift correspondingly.

Overlaying, facilitating and informing material commitments are discursive commitments. These may be specifically directed towards an existing system configuration (e.g. engineering discourses around electricity systems). Alternatively, wider discourses may support particular constituent socio-technical structures and practices (e.g. neo-liberalism and electricity markets). These discursive commitments may be articulated for varying substantive, normative or instrumental reasons (Stirling, 2005). Whatever form they take, they play an important role in co-ordinating action amongst governance actors. In this regard, it is desirable that discursive commitments display a degree of interpretive flexibility, such as to maximize the engagement, recruitment or assimilation of a sufficient range of actors (Hajer, 1995).

Like the processes of social appraisal with which we wish to contrast them, discursive commitments have obvious epistemic qualities. Here, the crucial difference between discursive commitments and social appraisal, hinges on the distinctively 'ontological' character of commitments. For example, when a government minister articulates the *specific merits or shortcomings of* a socio-technical configuration like nuclear power, then the epistemic aspects of this intervention render it a contribution to wider discourses of social appraisal. Where the minister's intervention has the effect of *asserting a position,* by contrast (for instance signalling confidence over future support for nuclear power), then the intervention has a more ontological form—as a tangible expression of commitment. The effect may then be to co-ordinate a cascade of resulting commitments on the part of other actors. In this way appraisal and commitment are mutually co-constitutive—and possibly present as aspects of the same intervention—but remain analytically distinct for our purposes.

Institutional and structural factors complicate this picture of material and discursive commitments. This can be approached by distinguishing the role of established commitments in system reproduction from new commitments to system change. Governance processes seeking to transform a socio-technical system are structurally constrained by historically established commitments embodied in infrastructures, networks, institutions, practices and discourses. The formation of commitments for socio-technical transitions must therefore work against these structures, and require strategic governance intervention. For instance, new imperatives may be introduced through constitutive commitments to new innovation activities or liability regimes or new actors may be enrolled through discursive commitments to sustainability.

Many actors are committed to precisely the features of the socio-technical system which governance is seeking to change. Such actors typically contribute to the reproduction of the socio-technical system and enjoy advantages from that position. Examples in the energy field include utility companies, regulators, consumer associations and capital goods developers. Some of the resources that these actors apply to the reproduction of system functions will be essential in realizing commitments for change. Actors more intensively involved than others in system reproduction enjoy quite powerful positions, benefit strongly from the *status quo,* and occupy important gate-keeping positions (Smith *et al.,* 2005). Established commitments are often reinforced (and entrenched) by past investments in supportive infrastructures (e.g. fuel supply networks) and institutions (e.g. energy market regulations). Whilst these investments stabilize commitments to the existing socio-technical system, they also represent a form of structured power, which must be overcome in the formation of commitments to more sustainable configurations.

As a system undergoes change towards more sustainable configurations, the socio-technical positioning of different actors and their commitments undergoes concomitant shifts. Thus, a renewable energy utility, whose position and voice was relatively marginal in the incumbent, centralized, fossil fuel energy system, becomes more central in a decentralizing, low carbon energy system. Conversely, incumbents unable to adapt will suffer declining influence and benefits (e.g. profitability). Such threats are an inevitable source of resistance. Actors in structurally powerful positions under the *status quo* are well placed to exercise strong influence. Stoker (1998, p. 22) noted how "in a governance relationship no one organization can easily command, although one organization can dominate a particular process of exchange" and, clearly, there will be biases against radical socio-technical change.

Appraisal

Against this background, appraisal emerges as the 'epistemic' corollary to the more 'ontological' notion of commitments. In other words, appraisal concerns 'ways of knowing' rather than 'ways of being' in the socio-technical system. Accordingly, appraisal is about producing substantive understandings, social learning and cultural meanings concerning the socio-technical system. It is on this basis that more concrete commitments are formed in wider governance. In these broad terms, then, appraisal involves a wide variety of discursive processes, institutional practices, disciplinary approaches and methodological tools. In contrast to conventional distinctions between different styles of appraisal—such as expert-analytic versus participatory-deliberative—this analysis highlights two important, cross-cutting distinctions. The first concerns the breadth of the *inputs* to appraisal: which may variously be 'broad' or 'narrow' in a number of different ways. The second concerns the way that the *outputs* of social appraisal serve to 'open up' or 'close down' the formation of discursive and material commitments in wider governance.

There are many dimensions to the breadth of appraisal inputs. These concern all aspects of whatever is held to constitute salient bodies of knowledge—extending from technical and financial to environmental, social and ethical considerations. Breadth includes the extent to which causal relationships are explored and represented, concerning successively less 'direct' or more 'complex' effects and implications. It includes the treatment or acknowledgement of associated uncertainties and ambiguities and the range of intervention options that are addressed (alternative technologies, measures, policies, practices and institutional frameworks). A further dimension of breadth concerns the variety of contending social, institutional or disciplinary perspectives that are involved in producing and deliberating the knowledge, learning and meanings in question. All else being equal, the broader the appraisal process, the more 'precautionary' the resulting picture of governance options, in the sense that greater attention is paid to the full range of associated uncertainties, contingencies, conditions and contexts (EEA, 2001).

Irrespective of whether inputs are 'broad' or narrow', the outputs of appraisal may equally be variously 'open' or 'closed' in form (Stirling, 2005). This focuses on the manner in which the consequences of appraisal are represented to wider processes of governance. Does this take the form of 'closure' around the merits of particular technologies or interventions? Or does it rather comprise a more reflexive 'opening up' of the contingencies, contexts, conditions or perspectives under which different possible technologies or governance interventions might alternatively be favoured? This is not a necessary corollary of breadth in the inputs to appraisal. A broad-based 'precautionary' appraisal process may yield normative grounds for the 'closing down' of commitments around a particular course of action. A relatively narrow technical expert appraisal, on the other hand, may (through procedures like sensitivity analysis or minority opinions) serve to 'open up' wider discourse, by highlighting the validity of a variety of contending judgements or interpretations.

In these terms, then, closing down in appraisal is about "defining the right questions, finding the priority issues, identifying the salient knowledges, recruiting the appropriate protagonists, adopting the most effective methods, highlighting the most likely outcomes and so determining the 'best' options" (Stirling, 2005, pp. 21–22). Opening up, by contrast, reveals to wider governance discourses the open-endedness, contingency and capacities for social agency in technology choice. "Instead of focusing on unitary prescriptive recommendations, appraisal poses alternative questions, focuses on neglected issues, includes marginalised perspectives, triangulates contending knowledges, tests sensitivities to different methods, considers ignored uncertainties, examines different possibilities and highlights new options" (Stirling, 2005, p. 22). In particular, by highlighting the way in which the picture yielded in appraisal is contingent on, and conditioned by, prior perspectives—and thus commitments—of different actors, an 'opening up' mode represents a greater degree of reflexivity.

This analytical distinction between interlinked epistemic processes of appraisal and the more ontological formation of commitments speaks to key issues in the 'reflexive governance' literature. Here, contributions use the term 'reflexivity' in a number of quite fundamentally different ways (Adam *et al.*, 2000; Beck, 1992; Giddens, 1990; Lash, 2001). In our view, reflexive governance is not simply a strategic orientation based around the idea that interventions generate unintended consequences that require governance to rethink and respond differently. This kind of literal reflex has dogged earlier forms of governance too. Recent literature does argue that unintended consequences are becoming more severe and disabling (Beck, 1992) and a corollary for some is that 'mastery' over socio-technical systems is nigh impossible (Latour, 2003, cited in Rip, 2006). Others suggest a pragmatic governance must recognize its own implication in unintended consequences and anticipate adaptive responses. This involves what Voβ *et al.* (2006) termed "second order reflexivity" in the sense that governance becomes aware of how certain forms of appraisal process—e.g. narrow inputs—can exacerbate unintended consequences, and so avoids those kinds of process by seeking more integrative and interdisciplinary knowledge, operating across a broader consideration of factors and taking longer-term and adaptive perspectives. An objectifying, 'governance on the outside' perspective (next section) can accommodate this kind of reflexivity. Whilst not disputing these aspirations, they do not reveal an important additional quality in *fully reflexive governance*, which is awareness of how appraisal and commitments condition, represent and recondition one another recursively, such that *social appraisal rarely closes down definitively upon a given socio-technical object*. 'Governance on the inside' develops strategies for accommodating this realization (which is not a new phenomenon—see Law & Urry, 2004; Scott, 1998). The article now turns to the two ideal-typical conceptualizations of governance, based upon different understandings of the relations between social appraisal and social commitment formation.

Governance on the Outside: Intervening in Socio-Technical Change

Having characterized the distinction between appraisal and commitment, a framework for analysing governance is now developed using two ideal-type conceptualizations based on contrasting appraisal–commitment relationships. This may contribute to understandings of dynamic tensions between managerial processes seeking to objectify the socio-technical and political processes constructing the socio-technical (Table 1), The first ideal-type—'governance on the outside'—considers governance arenas as conceptually apart from socio-technical objects. The second ideal-type—'governance on the inside'—conceptualizes governance as co-constituting the socio-technical. In practice, and in much of the literature, both ideal-types have a presence.

Intrinsic to a 'governance on the outside' view is a socio-technical object whose unique operation, boundaries and consequences can become known to governance through appraisal, and that can be predictably altered through a reordering and redeployment of commitments. Nearly all accounts of governance recognize that this requires negotiation across different actor perspectives. However, 'governance on the outside' still perceives this as something that can be negotiated with reference to a commonly held, unique and discrete socio-technical object. In other words, the domains of environmental sustainability, the socio-technical system and governance itself are each conceptualized as essentially separate and knowable in their own right. Only in this way can governance processes appear sufficiently rational and synoptic—providing self-evident frameworks for identifying an objectively 'best' plan for intervening in the socio-technical system.

Figure 1 is structured schematically around the relationship between appraisal and commitment developed above. The first thing to notice is how governance seeks to close down the outputs of appraisal around the socio-technical object. The inputs to appraisal can be broad (as defined in the last section), but because a definitively knowable socio-technical object is presumed to be 'out there', then appraisal can be expected to identify this and yield as 'outputs' a picture of the 'best' options. As such, 'governance on the outside' presumes a single episteme and may therefore be reflective (depending on the breadth of inputs) but not reflexive (relating to openness of outputs). The multiple perspectives and knowledges held by relevant actors furnish individual pieces of a single overall jigsaw. Broadening the inputs to appraisal creates a more complete picture, with each perspective contributing slightly more information. As such, the meta-governance task is to ensure that relevant stakeholders are consulted appropriately, so that governance can close down around the best option identified in appraisal.

Simplistic use of terms like "sound science" (Blair, 2000; 2003) and "evidence-based policy" (DEFRA, 2004) imply this perspective. Sustainability indicators are treated literally—as 'metrics' of the socio-technical object, its impact on the environmental system, and for recording progress. Appraisal is episodic and learning is orientated towards checking the efficacy of the committed interventions in relation

Table 1. Governance perspectives and implications for processes of appraisal and commitment formation

Governance function	Governance perspective	
	'Governance on the outside' external intervention by governance subject in socio-technical object	'Governance on the inside' internal co-constituting of governance and socio-technical subjects
Appraisal	• Broadening-out the inputs to appraisal/extended reflection • Scoping a particular sustainability problem/goal • Aggregating 'relevant' actor perspectives • Sustainability indicators treated as metrics • Drive to objectify the socio-technical object • Informs formation of commitments • Analysis and deliberation over the 'best option(s)' • first-order learning: effectiveness of appraisal/intervention	• Opening-up the outputs of appraisal/pluralistic reflexivity • Accepting contested nature of sustainability • Exploring different actor framings • Sustainability indicators treated as heuristics • System ambiguity accepted • Empowering deliberation over commitments • Incommensurable perspectives, conditional and situated options • second-order learning: consequences of different framings
	Clear distinction between appraisal and commitment in governance stages.	*Reflexive interaction between appraisal and commitment processes in governance.*
Commitment	• Appraisal determines commitment formation • Managing governance interventions • Legitimacy derives from objectivity or authority of appraisal • Concentration and uniformity of commitment • Aversion to failure • Unilinear, unidimensional and discrete interventions • Episodic and isolated commitment making • Interventions seen in terms of function	• Appraisal conditionally informs commitment formation • Closure through wider political discourse • Legitimacy is negotiated through governance • Ensuring strategic diversity, resilience and robustness • Irony and social learning • Multilinear, multidimensional flexible repertoires • Acknowledges persistent (re)negotiatability • Interventions seen as power laden
Attitude to governance	*Largely instrumental managerial function*	*Fundamentally engaged political process*

to the particular closed commitments, monitoring subsequent changes to the socio-technical object, and recalibrating outputs accordingly.

Since closure is reached in appraisal, the implications for reconfiguring commitments is relatively straightforward. Appraisal not only reveals the socio-technical object and courses of action, but there is a clear distinction between established and new commitments. By integrating sufficiently broad perspectives, evidence attains an objective legitimacy free of existing commitments, such that it automatically entrains and reshapes the sustainability commitments required of

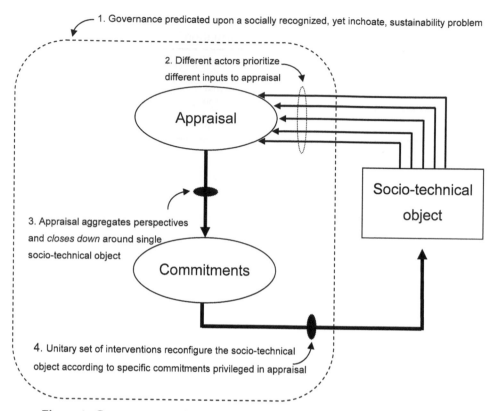

Figure 1. Governance on the outside: intervening in the socio-technical object

different actors. Commitment formation follows on deterministically from appraisal, and interventions have to realign established commitments into forms identified by the appraisal.

Thus, actors not only contribute to the objective appraisal of the system, but
also adhere to the intervention recommendations arising from that appraisal by
shifting commitments accordingly and reconfiguring the way they contribute to
the reproduction of the socio-technical system. This perspective is functionalist
in the sense that all actors respond to the imperative of the system overall and,
if that imperative becomes more inclusive of sustainability criteria, then actors
will redefine and remake their commitments accordingly. Clearly, this managerial
task has a political dimension, but ultimately it remains 'managerial' because the
politics is about strategies for persuading, cajoling and forcing actors to bring their
commitments into line with the closure reached in appraisal. Governance 'fails'
either because the appraisal is subsequently recognized to have been inaccurate,
or because insufficient actors fall into line.

Dilemmas Confronting Governance on the Outside

A 'governance on the outside' position relates to a positivist position that is difficult to maintain (Latour, 2004). In a complex and dynamic world, closure around
appraisal will always be provisional. Pervasive uncertainties and surprises of one
sort or another will arise, leaving space for competing (re)interpretations. Even

under conditions of broad consensus, the managerial challenge can appear daunting. Difficulties in co-ordinating between processes over different scales, compounded by mismatches between emergent structures at different levels and disruptions by contingent events, reinforce Kooiman's point about governance forever trying to 'catch up'. Some degree of governance 'failure' appears inevitable (Jessop, 2003). Nevertheless, this perspective can have analytical advantages, to the extent that it reveals in stark form the governance processes necessary to objectify and manage socio-technical systems—even though political realities mean those externalizing processes are inherently problematic.

Challenges are not confined to the incomplete integration of inputs in closing-down appraisal, nor the insufficiency of power required to bring actor commitments into line. Both of these could be overcome by doing 'governance on the outside' better (e.g. more computational power in appraisal, greater legitimate authority over commitment formation). Indeed, theories and practices that lean towards 'governance on the outside' advocate both these measures. However, in analysing relations between appraisal and commitments, one can begin to understand why 'governance on the outside' will always be problematic. More of the same will not work.

One fundamental difficulty confronting 'governance on the outside' is the essentially contested nature of sustainability (Jacobs, 1999). Decades of work in the field of social choice has shown that there cannot—either in principle or practice—be any definitive means to integrate divergent perspectives, interests and preferences, such as to yield a single coherent ordering of socio-technical (or other policy) options (Arrow, 1963; MacKay, 1980). Such managerial aspirations are confounded by the incommensurable dimensions of socio-technical performance, strongly divergent socio-political interests and perspectives (Brown *et al.*, 2000), recursive interrelationships between commitments and appraisal, and the profound and ever-present exposure to surprise (Wynne, 1992). Further intractable issues are raised concerning the role of power (Lukes, 2005) and the nature of effective social deliberation (Habermas, 1996; Munton, 2003) in the formation of 'sustainability'.

Others interpret this view to undermine notions of 'public interest' in sustainability (Meadowcroft, 2005). Yet, this is not necessarily so. That there can be no guarantee of definitive specifications of 'public interest', does not preclude the resolving of partial characterizations or rank orderings, that some possibilities might unambiguously be excluded or that or that robust areas of consensus might not arise from time to time. The real message is to concentrate on legitimate and effective means to social deliberation and learning, and on the crucial role of nurturing plurality, reversibility and sustained dissent in any provisional constructions of the 'public interest'. This, in turn, raises challenges over the maintaining of requisite diversity and resilience in social commitments to socio-technical trajectories (Brooks, 1986; Stirling, 2007). As social priorities shift; knowledge develops; and power relations are challenged, so there exists a premium on those trajectories that are more adaptable, or from which it is easier to withdraw (Stirling, 2003; Wynne, 1992). The resulting complexities mean that governance processes typically remain susceptible to the ever-present possibility of spontaneous, dissenting political discourses opening up new terrain. This leaves the appraisal bedrock of 'governance on the outside' susceptible to significant upheavals.

Against this background, Voβ *et al.* (2006, p. 427) encouraged governance "to establish a setting that is appropriate for the relevant problem. In short, the interaction space needs to be congruent with the problem space". In addition to the sustainability 'problem space' being open to different framings, then, so is the socio-technical problem space also ambiguous and uncertain. Amongst analysts there is ambiguity over the most effective level of empirical application of the socio-technical concept (Berkhout *et al.*, 2005). Different participants in a system of socio-technical practice hold contrasting mappings or framings of the 'system' and their role in it (e.g. appropriate boundaries, key causes of sustainability problems, reconfigurations that will resolve the problem, interrelations between system components). Systems are open to subjective interpretation and inter-subjective construction. As Voβ *et al.* (2006, p. 423) put it, this "means that the agent of governance gets displaced from its Archimedean point, outside of the developmental context. Instrumental rationalization and steering are not applicable under these conditions". A common framework—if any—will be the emergent product of negotiations between the different actor framings (and the commitments underpinning each)—and so itself likely partly contingent and path-dependent.

Taking the electricity socio-technical system as an example, the 'mental maps' of householders with respect to the system providing energy services need not correspond with that of the analyst, governance agencies or other members of the 'system'. Even an actor more actively and intensively involved in the reproduction of a socio-technical system, such as an energy utility company, need not have a comprehensive map of their position in the wider system. The energy utility is primarily concerned with customer markets, competitive generating technology, immediate infrastructures, shareholder value and, possibly, social reputation. It maps its energy system accordingly. Governance for sustainability concerns the utility to the extent that it might reframe the context for business operations. If the utility chooses to participate in sustainability governance processes then it, like the householder (or consumer association), will bring to negotiations a particular framing of the socio-technical, associated problems, and plausible solutions.

It is important to stress the co-evolutionary development of these frames. Contrary to the 'external' view, it is not that each actor holds a fragment of a whole, such that all mental maps can be stitched together to reveal the overall picture. Actors can hold incommensurable framings, rendering problematic any straightforward aggregation. Each actor's framing co-evolves with the framings of others. As a result, governance necessarily involves interplays between different framings of the 'system' and their associated concerns and priorities. These interactions mean actors constitute one another's room for manoeuvre (Rip, 2006). Each framing is associated with different forms of commitment and various degrees of power and agency to effect change. Some selection processes are more 'co-' than others in 'co-evolution'; and the logics within one system framing can have an overriding dominance compared to the logics of other system framings (Jessop, 2002). If some form of socio-technical construct is agreed (or imposed) collectively, say around energy, and commitments reformed accordingly, then each actor's original framing fits the changed understanding of the situation even more imperfectly than before. Everyone must learn anew, together. 'Governance on the outside' will find its socio-technical object quite elusive.

Sustainability governance can consequently be characterized as the emergent outcome of attempts by different coalitions of actors to set priorities and conceptualize the socio-technical system and its concomitant sustainability problems. Some framings will accommodate broader coalitions of actors than others; some will have enrolled support from powerful actors in structurally privileged positions. If governance is to work, then participants must identify and negotiate overlaps between their framings as a mutual basis for interaction. In this regard, Jessop (2003, p. 115) noted how provision for broad deliberation not only becomes a desirable feature of governance, but reflexively how it "will affect in turn the definition of the objects of governance and, insofar as governance practices help to constitute these objects, it will also transform the social world that is being governed". Such constituting practices are highly political, which brings us to the 'governance on the inside' perspective.

Governance on the Inside: Co-Constituting Governance and Socio-Technical Subjects

'Governance on the inside' acknowledges the contestability and interpretative challenge identified above and seeks an accommodation with it. It is a reflexive mode of governance since it explicitly recognizes at the outset that there are multiple ways of knowing the socio-technical system, each valid in its own way, but with different implications for the way governance engages with and affects it. The image here is of governance actors and processes as inseparable, pervasive and partly co-constitutive of internal features of the socio-technical system itself. Figure 2 summarizes schematically how 'governance on the inside' operates in terms of relations between the socio-technical, appraisal and commitments.

As with 'governance on the outside', an internal perspective seeks to broaden the inputs to appraisal, but rather than reflecting and aggregating multiple perspectives around a shared framing (the socio-technical object), the conceptualization here accepts the possibility of incommensurable framings. There are multiple epistemes. Acknowledging how this renders socio-technical sustainability ambiguous and indeterminate calls for appraisal to be deliberately and pluralistically reflexive. Commitments to different framing conditions prioritize, constrain and shape different salient socio-technical features for appraisal, as well as suggesting alternative methodologies for attaining knowledge about those features, thereby engendering distinct understandings of the system, its sustainability and 'optimum' intervention strategies for change. The governance task becomes: constituting (rather than inheriting) networks; testing (rather than assuming) legitimacy; negotiating (rather than imposing) expertise; challenging (rather than accommodating) power; and exercising a facilitating authority based on pluralism rather than objective neutrality.

An important consequence of this perspective is that (though appraisal certainly informs), closure is not reached through objectification, but through negotiated commitment formation. Commitments precede, mediate and are partly formed through appraisal, but important aspects of commitment formation also arise through the diverse normative goals of broader political discourse. There is a reflexive circularity, in the sense that established commitments and normative political discourses are already informing the multiple and incommensurable framings being grappled with by the appraisal function of governance. Here, an 'ironic' (Jessop, 2003) separation is maintained between appraisal and commitment, in order to establish an analytical purchase on the complexities. This contrasts with attempts at more literal separation under an 'external' view.

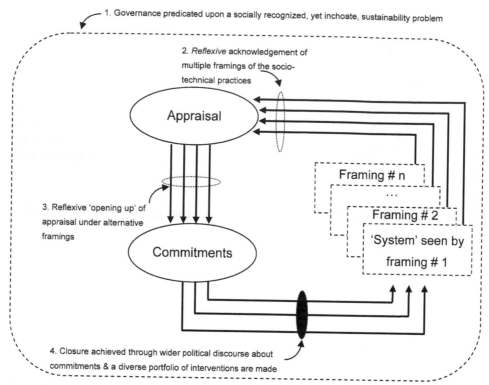

Figure 2. Governance on the inside: co-constituting of governance and socio-technical subjects

This also contrasts with an emphasis in 'governance on the outside' on first-order learning about the instrumental efficacy of interventions, in that an internal conceptualization of governance extends this to include more normative lessons about the socio-technical consequences of different framings (Voβ & Kemp, 2006).

Clearly, closure under the internal co-constitutive account is a much more complex process than the elicitation of unitary outcomes in appraisal. The forming of social commitments to particular technologies is understood in a more conditional, temporary, diffuse and reversible fashion than is suggested by the discrete notion of 'decisions' (Wynne, 1992). Accordingly, rather than the monolithic optimizing strategies associated with external governance, internal governance recognizes the importance of strategies for 'closure' that build in qualities of flexibility, diversity, resilience and robustness. In other words, the closure that still takes place is as much ontological (embodied in commitments) as it is epistemic (embedded more exclusively in appraisal). Figure 2 illustrates this strategy with governance committing to a diversity of socio-technical options.

In 'governance on the inside', processes of engagement, dialogue and deliberation require explicit and careful attention to questions of power, authority, consent, dissent and, above all, legitimacy. In particular, governance must undertake an open and inclusive normative evaluation of "the correctness of its procedures, the justification for its decisions, and the fairness with which it treats its subjects" (Grafstein, 1981, p. 456 quoted in Beetham, 1991, p. 10). Legitimacy has a double role here. First, there is the legitimacy of the governance deliberations themselves. Efforts are made to ensure key uncertainties are acknowledged, different assumptions and frameworks are rendered transparent, the plurality of social values is debated and different material interests are addressed. Secondly, there is the legitimacy of governance constructions of the socio-technical system. The legitimacy of established practices and ideas about sustainability must be questioned in governance deliberations and new sources of legitimacy explored for more sustainable options.

Dilemmas Confronting Governance on the Inside

Sources of governance failure are more complex than with governance on the outside. Governance failure on the 'inside' arises through limitations in the degree of reflexivity that is actually achieved (Jessop, 2003). Idealized symmetrical social partnerships, helpful for deliberations in 'governance on the inside', fail to be borne out in practice. This problem is compounded by the perennial insufficiency of reflexivity in the way governance processes engage with the complex systems in which they are embedded, and which they are trying to shape.

'Governance on the inside' must confront a number of dilemmas. A particular challenge arises in the close relationship between 'sustainability governance' and 'reflexive governance' (Voβ et al., 2006). The objectives of sustainability (broad and sometimes ambiguous as they are) do not have a monopoly of salience in the understanding and directing of socio-technical change. Sustainability governance 'merely' attempts to advance *sustainable* socio-technical change arguments and initiatives. A sustainability governance arrangement will therefore play a constitutive role to the extent that it becomes *the* focusing process for negotiating strategic socio-technical changes being considered in other arenas (e.g. markets, board-rooms, regulatory agencies, government ministries). There may, for example, be separate governance arrangements aimed at boosting the international competitiveness of firms or sectors that occupy overlapping socio-technical territories, but operate different sets of criteria and activities. After all, existing institutions and governance arrangements not concerned with sustainability questions are also intrinsic elements of the socio-technical system and will need to be addressed in any governance moves for sustainability.

Furthermore, the relatively institutionalized deliberations and decisions taken in formal governance settings are not the only arena in which socio-technical practices and problems are considered and articulated. Governance takes place within a wider and more spontaneous political discourse that can, from time to time, disrupt and penetrate more formalized governance deliberations and activities (Hajer, 1995; Torgerson, 2003). As an example, the ebb and flow of different narratives within political discourse on energy (e.g. environment, security, poverty, dependency, energy 'gaps', liberalization) have been reflected in shifting patterns of support for different socio-technical practices in energy governance. The performance of these different practices is reconsidered against the newly salient criteria or concerns in political discourse (e.g. the rise, fall and current attempted revival of nuclear energy).

In addition, care must be taken to avoid the relativistic trap. To acknowledge that the sheer complexity of natural and socio-technical systems introduces a degree of indeterminacy, does not imply that 'anything goes' (Cilliers, 2005; Grin, 2006; Stirling, 2006). Material structures and institutional processes, however complex, still constrain the way sustainability governance arrangements can interpret and 'construct' an agenda for a socio-technical system. There will be elements of the day-to-day operation and development of technological practices associated with the socio-technical system that offer limited interpretative flexibility and resist assimilation to the priorities of sustainability governance. Power relations and established structures—as well as the inherent properties of technical artefacts and natural environments—limit the diversity of framings and social constructions of socio-technical systems that are available to sustainability governance. The argument is not over whether 'closure' will necessarily—or desirably—occur. Rather, it concerns the locus, form and degree of this social closure.

In other words, the conceptual positioning of governance as an internal feature of socio-technical systems does not render these 'internal' governance processes into the sole constituting forces. Rather, governance activities connect and interpret broader socio-technical realities and wider political discourses and provide an important focal site. So the *co*-constitutive role of governance is as a deliberative site shaped by, and shaping, socio-technical systems and political discourses.

As with 'success', so is it important to recognize the limits and conditionalities attached to notions of 'failure'. Just as solutions can seldom be 'optimal', so are governance failures rarely complete. Incompleteness, insufficiency or divergence from initial aims is usually qualified by mitigating factors. Moreover, satisficing must not be interpreted conservatively and follow only incremental approaches. More ambitious and radical paths can be realized. The crucial point is that governance retains faculties for reflexivity, flexibility and irony in respect of failure as much as success. This ironic stance derives from recognition of how one is reflexively implicated in the imperfect social construction of a socio-technical 'object' and sustainability 'goal'. It demands greater humility over limits and fallibility of both analysis and deliberation, whilst retaining optimism over the efficacy of action (Jessop, 2003). In these terms, it is better to acknowledge how power relations help shape and curtail deliberation, than to pretend such distortions do not exist (Meadowcroft, 1998).

A second strategy worth highlighting, this time exclusively under 'governance on the inside', is deliberately to cultivate a "flexible repertoire of responses so that strategies and tactics can be combined in order to reduce the likelihood of failure and to modify their balance in the face of failure and turbulence in the policy environment" (Jessop, 2003, p. 107). From a narrow (economic) and short-term perspective, this may look like inefficient redundancy. But under broader and longer-term perspectives, such flexibility is an essential response to dynamic and uncertain environments. These general governance injunctions resonate with the wider literature on adaptive capacity, resilience and robustness (Folke *et al.*, 2004).

Governing Under Conditions of Ambivalence, Uncertainty and Distributed Power

This paper has developed a schematic dichotomy of sustainable socio-technical governance. 'Governance on the outside' is considered as *external* intervention acting upon an effectively separate socio-technical object: appraising the object, forming commitments accordingly and monitoring progress. Tasks are essentially instrumental and managerial. 'Governance on the inside' is seen as *internally* co-constituting the socio-technical system—conditioning and itself conditioned by the relationships, practices, problems and understandings which it seeks to steer. Tasks are consequently characterized as more political than managerial. Implications arising from these two ideal-typical conceptualizations are summarized in Table 1. Each suggests different strategies for realizing sustainable socio-technical transitions. Correspondingly, each approaches questions of ambivalence, uncertainty and power relations very differently.

'Governance on the outside', owing to its fundamental objectifying drive, considers ambivalence and uncertainty as temporary conditions. Uncertainties are 'closed down' through 'better appraisals'. Successive appraisals will, over time, feed lessons back into the continual (re)formation of commitments towards sustainability, progressively clarifying the ultimate object (a sustainable socio-technical system). Just as uncertainty is subject to reduction in this way, so too is any ambivalence on the part of contending actors, concerning the characterizing and prioritizing of different aspects of sustainability. Even if unclear at the outset, 'optimum' solutions and pathways are expected to become evident in due course.

The question of power in 'governance on the outside' is relevant in so far as governance processes need sufficient power to bend and co-ordinate actor commitments to the findings of appraisal. As understandings about the socio-technical system improve (uncertainty reduced through learning), and as the superior performance of certain sustainable socio-technical practices becomes more evident (ambivalence overcome through converging experience), so the power of argument is enhanced and the need for coercive power reduced. Managing ambivalence, uncertainty and power is undertaken by following strategies in the left-hand column of Table 1.

'Governance on the inside' sees things very differently. Uncertainties are transformed rather than reduced, since reductions under one framing merely pose more questions under alternative framings. Ambivalences prevail over how governance should best cut into and simplify the sheer complexities of socio-technical systems and dynamic relations with complex natural systems. Power relations are recognized to pervade the negotiation of governance constructs. Ever more accurate appraisal is unlikely to see the progressive evaporation of such power relations. Rather, such relations must be rendered transparent and subject to deliberation and scrutiny. Negotiations of ambiguity, uncertainty and power are undertaken through strategies in the right-hand column of Table 1. None of these serve to resolve ambiguity, uncertainty or concentrations of power. Rather they provide means for living with them.

The overriding strategy in the left-hand column is to objectify. The overriding strategy in the right-hand column is to enhance reflexivity. The question is, do our two types always contradict one another, or can they be complementary? The short answer is that they are inherently in tension. Each adopts very different approaches to the building of pathways to sustainability—with potentially crucial implications for the natures of these pathways themselves. A longer answer is more qualified, considering possible accommodations where governance is viewed over time. In reality, both types are present simultaneously to varying degrees. Indeed, the resulting tensions may play a key role in driving the unfolding dynamics of governance itself.

Thus, analysis can track (theoretical and practical) developments in socio-technical governance as they move around the cells in Table 1. If, for example, governance innovates appraisal procedures that open up the outputs (top right of Table 1), but still anticipates appraisal as determining the necessary commitment formations for sustainability (bottom left of Table 1), then this contradiction will soon become apparent in disputes over the basis of the course of action taken. Conversely, appraisals purporting to be 'objective' (top left cell in Table 1), but whose narrow framing actually corresponds tightly to prior sectional commitments, will also meet with difficulties.

Two recent examples in the UK serve to illustrate these tensions. The limited effect of the government-launched *GM Nation* debate in the UK illuminates the first pattern. Here, high expectations were created over an unprecedented process of broad and open engagement—effectively articulating in high-level appraisal perspectives that had hitherto been marginalized in earlier governance processes on this issue (DEFRA, 2003). Yet, the resulting effects on tangible policy commitments were minimal, accompanied by official attempts to discredit the basis and authority of the appraisal exercise itself (Grove-White, 2006). The rapid succession of formal government 'energy reviews' conducted in the UK between 2002 and 2004, on the other hand, illustrate the second pattern. Here, the first ostensibly 'objective' appraisal exercise concluded that renewable energy and energy efficiency should form the basis for low carbon energy strategies (SU, 2002). On effectively repeating the process out of concern that nuclear power had been unduly neglected (Adams, 2002), the conclusion in a subsequent White Paper was none the less that the nuclear option remained "unattractive" (DTI, 2003). Only on a quickly convened third attempt, did a further formal government 'Energy Review' manage effectively to 'trump' the earlier negative conclusions (DTI, 2006). Yet this final outcome was itself then overturned under Judicial Review on the grounds of an inappropriately constrained process of appraisal (Woodward, 2007). The revealing response on the part of the prime minister to the effect that this outcome will affect the consultation, but "won't affect the policy at all" (BBC, 2007) could hardly provide a more stark illustration of the practical distinction between appraisal and commitment!

More detailed case studies, elaborating on sketched examples such as those above, are needed to test and develop the scheme provided in this paper. Tendencies towards objectification and reflexivity in these governance histories need mapping, the coalitions who are seeking to objectify or introduce reflexivity need studying (bringing actors back in), as well as analysing the processes which bring the two tendencies into contact, and how one becomes undermined whilst the other comes to the fore, and when this becomes destabilized again.

Current attempts by theorists to formulate 'reflexive governance' can be seen as a conceptual move to the right-hand column of Table 1. Any difficulties in these attempts arise because they retain some of the managerial aspirations of governance in the left-hand column (see, for example, the edited volume by Voß *et al.* (2006)). These theorists suggest, depending on the context and perspective, it may or may not be reasonable to argue that 'framing politics' must cease at some point, and governance must arrive at some manageable, objectifying closure.

Even with the benefit of initial understanding and further learning, an 'eternal tension' will remain between the need to fix an approach in order to do something here and now and perhaps make a difference, and the still insufficient understanding of what might happen which makes it difficult to 'fix' the appropriate approach (Rip, 2006).

However imperfect or provisional they may be, commitments will (of course) be formed. Whether consensual, majoritarian, elitist, or to meet sectional interests, pragmatic 'decisions' must be made (or at least be seen to be made). This analysis reminds reflexive governance theorists why any (managerial) 'implementation' will always be provisional and, indeed, why 'decisions' have to be put into a broader historical context. In both cases (but for different reasons), governance must substitute hubristic aspirations to optimal solutions with more modest 'satisficing' strategies aimed at "acceptable outcomes" (Simon, 1983). In practice, satisficing happens to varying degrees in virtually all governance settings (even if outward-facing representations present decisions as optimal). A more open and transparent satisficing strategy requires regular self-critical re-assessment of the extent to which these strategies are delivering the outcomes desired under different perspectives.

Conclusions

The central argument in this paper is that the manner in which governance processes may realize[1] sustainable socio-technical systems depends on the general way in which each is conceptualized in relation to the other. Too often in the socio-technical systems literature, these conceptualizations remain implicit and confused. In this paper two contrasting ideal-typical conceptualizations provide a heuristic for studying processes of objectification and reflexivity in governance more systematically. Analysing governance processes using the heuristic here seeks to understand the tense combinations of (contradictory) strategies exemplifying governance on the outside and inside (Table 1), and how these propel policy developments.

The argument is not that conceptual distancing between socio-technical governance subjects and objects is always necessarily unhelpful or wrong. The point is rather that interlinkages between socio-technical appraisal and commitments should become more reflexive. In particular, there needs to be greater appreciation of the *internal* loci of governance processes within the socio-technical systems themselves, and of the necessarily more 'open' role of appraisal under these conditions. In contrast to governance conceptualized outside the system, positioning governance inside the negotiation of socio-technical change requires processes for opening up debate and revealing technology's inherently political nature. In short, we need to move from a view of 'steering as management' to an understanding of 'steering as politics'.

In the end, sustainable development is an open-ended political process. The artificial separation of a governance 'subject' from an unsustainable socio-technical 'object' may provide some convenient simplifications for governance functions, but it quickly becomes apparent that this is just an expedient closure. Despite managerial attempts to elide ambiguities, obscure uncertainties and exclude dissent, neglected complexities have a habit of re-emerging in ever more unforgiving ways. At the same time, however, there will always be limitations to reflexivity and consequent needs (in certain governance settings, at least provisionally) to objectify and fix the socio-technical. In practice, therefore, governance both will (and should) move between the two poles established by our ideal-types of governance on the 'outside' and 'inside'. The key challenge is to acknowledge this dynamic and learn from these recursive movements between reflexivity and objectification.

Acknowledgement

Earlier versions of this paper were presented to various audiences and the authors are grateful to them for their feedback. Particular thanks go to to Richard Cowell, Andrew Flynn, René Kemp, Joseph Murphy, Jens Newig, Jan-Peter Voß and Jim Watson for helpful comments on earlier drafts of the paper. The usual disclaimers apply.

Note

1. The term 'realize' is intended in both senses here: to come to understand ones interests; and to try and fulfil those interests (after Byrne, 1998).

References

Adam, B., Beck, U. & van Loon, J. (2000) *The Risk Society and Beyond: critical issues for social theory* (London: Sage).
Adams, R. (2002) City Diary, *Guardian*, 13 December.
Arrow, K. (1963) *Social Choice and Individual Values* (New Haven: Yale University Press).
BBC (2007) BBC News, *Blair defiant over nuclear plans*, Thursday, 15 February. Available at http://news.bbc.co.uk/1/hi/uk_politics/6366725.stm (accessed 27 April 2007).
Beck, U. (1992) *Risk Society* (London: Sage).
Beetham, D. (1991) *The Legitimation of Power* (London: Macmillan).
Berkhout, F. (2002) Technological regimes, path dependency and the environment, *Global Environmental Change*, 12(1), pp. 1–4.
Berkhout, F., Smith, A. & Stirling, A. (2005) Socio-technical regimes and transition contexts, in: Boelie Elzen, Frank Geels & Ken Green (Eds) *System Innovation and the Transition to Sustainability: Theory, Evidence and Policy*, pp. 00–01 (Camberley: Edward Elgar Publishing).
Bezembinder, T. (1989) Social Choice Theory and Practice, in: C. Vlek & T. Cvetkovitch (Eds) *Social Decision Methodology for Technological Projects*, pp. 00–01 (Dordrecht: Kluwer).
Blair, T. (2000) *Speech Delivered by the UK Prime Minister to The European Bioscience Conference*, London, Monday, 20 November. Available at http://www.monsanto.co.uk/news/ukshowlib.phtml?uid=4104 (accessed 15 April 2005).
Blair, T. (2003) Response to Questions, 10 November, *Hansard*, Column 14 W, London. Available at http://www.publications.parliament.uk/pa/cm200203/cmhansrd/vo031110/text/31110w04.htm#31110w04.html_spmin1 (accessed 15 April 2005).
Brand, R. (2005) *Synchronising science and technology with human behaviour* (London: Earthscan).
Brooks, H. (1986) *The Typology of Surprises in Technology, Institution and Development*, in: W.C. Clark & R. E. Munn (Eds) Sustainable Development of the Biosphere (Cambridge: Cambridge University Press).
Brown, N., Rappert, B. & Webster, A. (Eds) (2000) *Contested Futures: A Sociology of Prospective Technoscience* (Aldershot: Ashgate).
Byrne, D. (1998) *Complexity and the Social Sciences* (London: Routledge).
Cilliers, P. (2005) Complexity, deconstruction and relativism, *Theory, Culture & Society*, 22(5), pp. 255–267.
DEFRA (2003) *GM Nation: findings of a public debate* (London: Department for Environment, Food and Rural Affairs). Available at http://www.gmnation.org.uk/ (accessed 22 August 2006).
DEFRA (2004) *Evidence and innovation: Defra's needs from the sciences over the next 10 years* (London: Department for Environment, Food and Rural Affairs). Available at http://www.defra.gov.uk/science/documents/forwardlook/ScienceForwardLook3rd.pdf (accessed 22 August 2006).
DTI (2003) *Energy White Paper: our energy future—creating a low carbon economy* (London: UK Department of Trade and Industry, HMSO). Available at http://www.dti.gov.uk/energy/ energy-policy/energy-white-paper/page21223.html (accessed 16 July 2006).

DTI (2006) *The Energy Challenge: report of the UK Government Energy Review* (London: UK Department of Trade and Industry). Available at http://www.dti.gov.uk/files/file31890.pdf (accessed 16 July 2006).

EEA (2001) *Late Lesson from Early Warnings: the precautionary principle 1898–2000* (Copenhagen: European Environment Agency).

Elzen, B., Geels, F. & Green, K. (2005) (Eds) *System Innovation and the Transition to Sustainability: Theory, Evidence and Policy* (Cheltenham: Edward Elgar).

Feenberg, A. (2002) *Transforming Technology: a critical theory revisited* (Oxford: Oxford University Press).

Folke, C., Carpenter, S., Walker, B., Scheffer, M., Elmqvist, T., Gunderson, L., Holling, C. S. (2004) Regime shifts, resilience, and biodiversity in ecosystem management, *Annual Review of Ecology Evolution and Systematics*, 35, pp. 557–581.

Geels, F. (2002) *Understanding the Dynamics of Technological Transitions* (Enschede: Twente University Press).

Giddens, A. (1990) *The Consequences of Modernity* (Stanford, CA: Stanford University Press).

Glasbergen, P. (1998) The question of environmental governance, in: P. Glasbergen (Ed.) *Co-operative Environmental Governance*, pp. 1–18 (Dordrecht: Kluwer).

Grafstein, R. (1981) The failure of Weber's concept of legitimacy, *Journal of Politics*, 43, pp. 456–472.

Grin, J. (2006) Reflexive modernisation as a governance issue, or: designing and shaping re-structuration, in: J. P. Voß, D. Bauknecht & R. Kemp (Eds) *Reflexive Governance for Sustainable Development*, pp. 57–81 (Cheltenham: Edward Elgar).

Grove-White, R. (2006) Britain's Genetically Modified Crop Contoversies: the agriculture and environment biotechnology commission and the negotiation of 'uncertainty', *Community Genetics*, 9, pp. 170–177.

Habermas, J. (1996) Popular sovereignty as procedure, in: J. Habermas, *Between Facts and Norms: Contributions to a Discourse Theory of Law and Democracy*, pp. 463–490 (Cambridge, MA: MIT Press).

Hajer, M. (1995) *The Politics of Environmental discourse: Ecological Modernisation and the Policy Process* (Oxford: Oxford University Press).

Hajer, M. & Wagenaar, H. (2003) Introduction, in: M. Hajer & Wagenaar, H. (Eds) *Deliberative Policy Analysis*, pp. 00–01 (Cambridge: Cambridge University Press).

Hertin, J. & Berkhout, F. (2003) Analysing institutional strategies for environmental policy integration: the case of EU enterprise policy, *Journal of Environmental Policy and Planning*, 5(1), pp. 39–56.

Hoff, J. (2003) A constructivist bottom-up approach to governance: the need for increased theoretical and methodological awareness in research, in: H. P. Bang (Ed.) *Governance as Social and Political Communication*, pp. 00–01 (Manchester: Manchester University Press).

Jacobs, M. (1999) Sustainable development as a contested concept, in: A. Dobson (Ed.) *Fairness and Futurity*, pp. 00–01 (London: Routledge).

Jessop, B. (2002) *The Future of the Capitalist State* (London: Polity Press).

Jessop, B. (2003) Governance and meta-governance: on reflexivity, requisite variety and requisite irony, in: H. P. Bang (Ed.) *Governance as Social and Political Communication*, pp. 00–01 (Manchester: Manchester University Press).

Kemp, R., Schot, J. & Hoogma, R. (1998) Regime shifts to sustainability through processes of niche formation: the approach of strategic niche management, *Technology Analysis and Strategic Management*, 10(2), pp. 175–195.

Kooiman, J. (1993) (Ed.) Governance and governability: using complexity, dynamics and diversity, in: J. Kooiman (Ed.) *Modern Governance*, pp. 35–48 (London: Sage).

Kooiman, J. (2003) *Governing as Governance* (London: Sage).

Lash, S. (2001) Technological Forms of Life, *Theory, Culture and Society*, 18(1), pp. 105–120.

Latour, B. (1996) *Aramis, or the love of technology* (Cambridge, MA: Harvard University Press).

Latour, B. (2003) Is re-modernizing occurring – and if so, how to prove it? A commentary on Ulrich Beck, *Theory, Culture and Society*, 20, pp. 35–48.

Latour, B. (2004) Politics of Nature: How to Bring the Sciences into Democracy (translated by Catherine Porter) (Cambridge, MA: Harvard University Press).

Law, J. & Urry, J. (2004) Enacting the social, *Economy and Society*, 33(3), pp. 390–410.

Leach, M., Scoones, I. & Wynne, B. (2005) *Introduction*, in: M. Leach, I. Scoones & B. Wynne (Eds) *Science, Citizenship and Globalisation*, pp. 00–01 (London: Zed).

Lukes, S. (2005) *Power: A Radical View*, 2nd edn (Basingstoke: Palgrave).

MacKay, A. (1980) *Arrow's Theorem: the paradox of social choice—a case study in the philosophy of economics* (New Haven: Yale University Press).

Marin, B. & Mayntz, R. (1991) (Eds) *Policy Networks* (Frankfurt: Campus Verlag).

Marsh, D. & Rhodes, R. A. W. (1992) (Eds) *Policy Networks in British Government* (Oxford: Oxford University Press).

Meadowcroft, J. (1998) Co-operative environmental regimes: a way forward?, in: P. Glasbergen (Ed.) *Co-operative Environmental Governance*, pp. 21–42 (Dordrecht: Kluwer).

Meadowcroft, J. (2005) Environmental political economy, technological transitions and the state, *New Political Economy*, 10(4), pp. 479–498.

Munton, R. (2003) Deliberative democracy and environmental decision-making, in: F. Berkhout, M. Leach & I. Scoones (Eds) *Negotiating Environmental Change*, pp. 109–136 (Cheltenham: Edward Elgar).

Nowotny, H., Scott, P. & Gibbons, M. (2001) *Rethinking Science: knowledge and the public in an age of uncertainty* (Cambridge: Polity).

Raven, R. P. J. M. (2006) Towards alternative trajectories? Reconfigurations in the Dutch electricity regime, *Research Policy*, 35, pp. 581–595.

Rhodes, R. A. W. (1997) *Understanding Governance* (Buckingham: Open University Press).

Rip, A. (2006) A co-evolutionary approach to reflexive governance – and its ironies, in: J. P. Voß, D. Bauknecht & R. Kemp (Eds) *Reflexive Governance for Sustainable Development*, pp.82–100 (Cheltenham: Edward Elgar).

Rip, A. & Kemp, R. (1998) Technological change, in: S. Rayner & E. L. Malone (Eds) *Human Choices and Climate Change Volume 2 – Resources and Technology*, pp. 00–01 (Columbus, Ohio: Battelle).

Rotmans, J. & Kemp, R. (2001) More evolution than revolution: transition management in public policy, *Foresight*, 3(1), pp. 15–31.

Russell, S. & Williams, R. (2002) Social shaping of technology: frameworks, findings and implications for policy with glossary of social shaping concepts, in: K. H. Sørensen & R. Williams (Eds) *Shaping Technology, Guiding Policy: Concepts, Spaces and Tools*, pp. 00–01 (Camberley: Edward Elgar).

Scott, J. C. (1998) *Seeing Like A State* (New Haven: Yale University Press).

Shove, E. (2003) *Comfort, Cleanliness and Convenience: the Social Organisation of Normality* (Oxford: Berg).

Simon, H. (1983) *Reason in Human Affairs* (Stanford, CA: Stanford University Press).

Smith, A. (2000) Advocacy coalitions and policy networks: explaining policy change and stability in industrial pollution control, *Environment and Planning C: Government and Policy*, 18(1), pp. 95–114.

Smith, A. (2006a) Green niches in sustainable development: the case of organic food in the UK, *Environment and Planning C*, 24, pp. 439–458.

Smith, A. (2006b) Governance lessons from green niches: the case of eco-housing, in: J. Murphy (Ed.) *Framing the Present, Shaping the Future: Contemporary Governance of Sustainable Technologies*, pp. 00–01 (London: Earthscan).

Smith, A. (2008) Translating sustainabilities between green niches and socio-technical regimes, *Technology Analysis & Strategic Management*, 00, pp. 00–01, in press.

Smith, A., Stirling, A. & Berkhout, F. (2005) The governance of sustainable socio-technical transitions, *Research Policy*, 34, pp. 1491–1510.

Stirling, A. (2003) Risk, Uncertainty and Precaution: some instrumental implications from the social sciences, in: F. Berkhout, M. Leach, I. Scoones (Eds) *Negotiating Change*, pp. 00–01 (Cheltenham: Edward Elgar).

Stirling, A. (2005) Opening Up or Closing Down: analysis, participation and power in the social appraisal of technology, in: M. Leach, I. Scoones & B. Wynne (Eds) *Science and citizens: globalization and the challenge of engagement*, pp. 218–231 (London: Zed).

Stirling, A. (2006) Precaution, foresight, sustainability: reflection and reflexivity in the governance of science and technology, in: J.-P. Voß, R. Kemp & D. Bauknecht (Eds) *Sustainability and Reflexive Governance*, pp. 00–01 (Cheltenham: Edward Elgar).

Stirling, A. (2007) A General Framework for Analysing Diversity in Science, Technology and Society, *Journal of the Royal Society Interface*, 4, (15) August pp. 707–719.

Stirling, A. (2008) Participation, power and progress in the social appraisal of technology, *Technology and Human Values*, in press (manuscript available from the author).

Stoker, G. (1998) Governance as theory: five propositions, *International Social Science Journal*, 155, pp. 17–28.

SU (2002) *The Energy Review*, Prime Minister's Strategy Unit, UK Cabinet Office, February. Available at http:// www.strategy.gov.uk/downloads/su/energy/TheEnergyReview.pdf (accessed 16 July 2006).

Torgerson, D. (2003) Democracy through policy discourse, in: M. A. Hajer & H. Wagenaar (Eds) *Deliberative Policy Analysis: Understanding Governance in the Network Society*, pp. 00–01 (Cambridge: Cambridge University Press).

Voβ, J. P. & Kemp, R. (2006) Sustainability and reflexive governance: an introduction, in: J. P. Voβ, D. Bauknecht & R. Kemp (Eds) *Reflexive Governance for Sustainable Development*, pp. 419–437 (Cheltenham: Edward Elgar).

Voβ, J. P., Kemp, R. & Bauknecht, D. (2006) Reflexive governance: a view on an emerging path, in: J. P. Voβ, D. Bauknecht & R. Kemp (Eds) *Reflexive Governance for Sustainable Development*, pp. 419–437 (Cheltenham: Edward Elgar).

Weber, M. & Hemmelskamp, J. (Eds) (2005) *Towards Environmental Innovation Systems* (Heidelberg: Springer).

Webler, T., Kastenholz, H. & Renn, O. (1995) Public participation in impact assessment: A social learning perspective, *Environmental Impact Assessment Review*, 15, pp. 443–463.

Woodward, W (2007) Judge deals blow to Blair's nuclear plans: court rules consultation on power stations was 'misleading and flawed', *Guardian*, Friday 16 February.

Wynne, B. (1992). Uncertainty and Environmental Learning: reconceiving science and policy in the preventive paradigm, *Global Environmental Change*, 2(2), pp. 111–127.

Wynne, B. (1995) Technology Assessment and Reflexive Social Learning: Observations from the Risk Field, in: A. Rip, T. Misa & J. Schot (Eds) *Managing Technology in Society*, pp. 19–36 (London: Pinter).

Wynne, B. (2001) Creating public alienation: expert cultures of risk and ethics on GMOs, *Science as Culture*, 10(4), pp. 445–481.

Index

Milton Keynes UK
Ingram Content Group UK Ltd.
UKHW051855071024
449327UK00025B/1973